Propulsion and Energy Systems

Edited by Konstantin Volkov

Published in London, United Kingdom

Propulsion and Energy Systems
http://dx.doi.org/10.5772/intechopen.1003413
Edited by Konstantin Volkov

Contributors
Anna Sukhenko, Fei Qin, Gulama Garip Alisher Ibrayev, Hiroki Watanabe, Ikkoh Funaki, Jialiang Wang, Konstantin Volkov, Nabil Bessanane, Nazgul Kaliyeva, Nursultan Doszhan, Tingting Jing, Xing Sun, Yuepeng Xin, Yuya Oshio, Zaure Rakisheva

First published in London, United Kingdom, 2025 by IntechOpen
IntechOpen is the global imprint of INTECHOPEN LIMITED, registered in England and Wales, registration number: 11086078, 167-169 Great Portland Street, London, W1W 5PF, United Kingdom

For EU product safety concerns: IN TECH d.o.o., Prolaz Marije Krucifikse Kozulić 3, 51000 Rijeka, Croatia, info@intechopen.com or visit our website at intechopen.com.

British Library Cataloguing-in-Publication Data
A catalogue record for this book is available from the British Library

Propulsion and Energy Systems
Edited by Konstantin Volkov
p. cm.
Print ISBN 978-0-85466-020-9
Online ISBN 978-0-85466-019-3
eBook (PDF) ISBN 978-0-85466-021-6

If disposing of this product, please recycle the paper responsibly.

Meet the editor

Dr. Volkov is a senior lecturer in fluid dynamics and thermodynamics at Kingston University (London, UK). He holds a Ph.D. in fluid mechanics. His areas of expertise cover multidisciplinary fields, ranging from the design and optimisation of energy systems to fundamental problems focused on the modelling and simulation of turbulent multi-phase flows. He is a Chartered Engineer, a member of the Institute of Physics and the Institution of Mechanical Engineers, a member of the Combustion Institute, and a Fellow of the Higher Education Academy.

Contents

Preface

Research in propulsion and energy systems is crucial for multiple sectors, including aerospace, transportation, defense, and power generation. Research leads to more fuel-efficient engines, which in turn reduce operational costs and environmental impact. In aviation and space, higher efficiency can increase payload capacity, range, or mission duration. Innovative propulsion technologies (like electric or hybrid-electric systems) help cut emissions and noise pollution. Research drives development in materials science, fluid dynamics, thermodynamics, and control systems. Research in propulsion and energy systems is foundational for technological progress, environmental sustainability, and national advancement. Continued investment in this field is essential to address 21st-century challenges and unlock future possibilities.

There remains a need to improve the safe operational envelope of propulsion systems while reducing environmental impacts. This has driven a demand to understand the fundamental physics of fluid flows and combustion mechanisms, as well as to promote clean and efficient engines. Improving the performance of propulsion and power combustion systems requires a detailed understanding of multiphase turbulent reacting flows. Further experimental and numerical investigations of various topics are needed in practical environments that better emulate engine conditions.

The book focuses on the development and improvement of methods and techniques for analysing and diagnosing the efficiency, operation, and maintenance of propulsion and energy systems, including fundamental research aspects of aeronautics/astronautics propulsion, power engineering, and practical applications. A particular focus of the book is fluid dynamics and heat transfer of internal flows and combustion processes associated with those devices.

The Introductory chapter gives a brief overview of propulsion technologies and the functioning of energy systems. Understanding fluid flow, heat transfer, and combustion physics is essential for safe operation (avoidance of stall, suppression of detonation), performance gains (thrust, efficiency), emissions compliance (meeting environmental regulations), and innovation (in hypersonic, hybrid-electric and green propulsion). Localised rich/lean pockets in a turbulent flame zone directly influence the formation of NOx, CO, soot and unburned hydrocarbons. Accurate modelling of these zones requires resolving both fluid turbulence and chemical kinetics.

In Chapter 2, the multiscale modelling and simulation of the regenerative cooling process of supercritical-pressure hydrocarbon fuels in ramjet engines is considered and discussed. A comprehensive and systematic review of the current research status on the thermal structure of advanced propulsion systems that utilise superalloys or carbon fibre-reinforced ceramic composites is presented. Then, the numerical modelling and simulation techniques describing the flow, heat transfer, thermal oxidation and pyrolytic coking phenomena of hydrocarbon fuels at supercritical pressures are discussed. Finally, multiscale numerical modelling and simulation techniques are employed for the thermal-mechanical analysis of composite materials.

Chapter 3 presents a numerical investigation of flow dynamics and heat transfer in rectangular mini-channels and pin-fin heatsinks, with a focus on diamond-shaped pin-fins arranged in a segregated, corrugated configuration. To accurately determine the heat transfer coefficient within this complex thermal system, key parameters (including mass flow rate, geometric dimensions, heat flux, and reference temperature) are thoroughly analysed. Special emphasis is placed on the method of calculating the reference temperature, as its precise estimation is crucial for optimising cooling performance. The study introduces a novel technique for accurately determining the temperature difference between the cooling fluid and the heatsink wall, thereby improving the approach for estimating the heat transfer coefficient. The numerical methodology is validated through experimental measurements, demonstrating that the proposed reference temperature calculation method provides a more accurate representation of the heat transfer coefficient.

Chapter 4 focuses on the numerical simulation of the flow of chemically reacting air at equilibrium conditions, considering high-temperature air effects. The results are compared to those obtained from a perfect gas model. The mathematical formulation of the high-temperature air model is provided, and it is verified and validated using computational fluid dynamics benchmark problems, such as the flow around a forward step and the flow around a sphere. The accuracy of numerical calculations of fluid quantities in high-speed flows is addressed. The flow patterns and shock wave structures computed using an inviscid fluid model, a model accounting for temperature-dependent viscosity, and a model accounting for turbulent effects are compared. Various options for simulating high-temperature processes in air are assessed. The shape and location of the bow shock wave, as well as the occurrence of numerical solution fluctuations behind the shock wave front, are highlighted, with an emphasis on the impact of different factors.

Chapter 5 provides a brief overview of Hall thruster research and development, focusing on its applications worldwide and efforts to achieve a highly fuel-efficient Hall thruster. The development goal is to expand Hall thruster technology for all-electric propulsion geostationary orbit satellites operating at multiple operational points to achieve high specific impulses and wide power ranges. These characteristics are important, particularly for space exploration. Until now, laboratory model thrusters and power processing units have been fabricated and tested using 6-kW-class thrusters. Additionally, extremely high-voltage Hall thrusters with specific impulses above 3,000 seconds have been experimentally investigated.

In Chapter 6, a numerical solution is derived for linearised equations relative to a perturbed orbit. A control algorithm is developed to ensure a sufficient level of formation stability in the presence of a disturbed reference orbit and various uncertainties based on robust control algorithms. The maximum control force required to maintain the satellite formation's configuration using these control algorithms is 7 mN. Since the satellite formation mission requires thrusters that consume up to 200 W, it is necessary to determine the types of propulsion systems that meet these requirements. Among the electric propulsion systems available on the market, such as ion engines and Hall thrusters, the specific impulse allows them to generate a maximum thrust of about 15 mN and a minimum of about 6.6 mN. Comparing the obtained thrust values with the 7 mN control force required to maintain the satellite configuration, it can be concluded that the developed mathematical framework can be successfully implemented in the satellite formation's control system in geostationary orbit, ensuring the control of satellite motion and their relative positions.

The book aims to advance the knowledge of propulsion and energy systems while fostering ongoing technological progress through computational analysis and experimental testing. The book promotes open discussion among research institutions, academia, and industry from around the globe on the research and development of enabling technologies. The book covers many aspects of theory and practice, delivering essential contributions and providing input and support to cooperative efforts.

Dr. Konstantin Volkov
Department of Mechanical Engineering,
Kingston University,
London, UK

Chapter 1

Introductory Chapter: Propulsion and Energy Systems

Konstantin Volkov

1. Introduction

Propulsion research refers to various technologies, techniques, and mechanisms used to move vehicles in aerospace industry and to generate or convert energy to support this movement. Energy systems generate, store, or convert energy for propulsion or onboard power. The aerospace industry faces escalating demands for efficiency, sustainability, and technological integration, with research trends spanning fuel optimization, emission reduction, hybridization, and AI-driven systems. Propulsion in high-speed vehicles such as supersonic jets, hypersonic aircraft, and high-speed vehicles demands advanced systems capable of delivering sustained thrust, thermal resistance, and aerodynamic stability at high velocities and temperatures [1].

2. Research directions

In a propulsion system, whether in jet engines, rockets, or gas turbines, key engine components work in a precise sequence to convert fuel energy into thrust or mechanical power. Air intake captures and conditions incoming air for the engine and slows down and compresses air (especially in supersonic aircraft) to match compressor requirements. Pumps and compressors compress air (in gas turbines) or pressurize propellants (in rocket engines) and increase pressure before combustion to improve thermal efficiency. Combustion chamber mixes fuel with air and burns it to release heat energy and generates high-temperature, high-pressure gas for downstream expansion. Turbine extracts energy from hot gases to drive compressors (in gas turbines) or turbopumps (in rockets). Nozzle converts thermal energy into kinetic energy to produce thrust and accelerates exhaust gases to supersonic speeds (in rocket and jet nozzles).

The development and improvement of methods and techniques for analysis and diagnostics of propulsion and energy systems is a critical area of research and engineering [2]. It focuses on optimizing performance, ensuring reliability, reducing downtime, and extending the life of complex systems in aviation and energy generation sectors. Improving the safe operation of propulsion systems while reducing environmental impacts is a focus of modern propulsion research and engineering. This involves balancing safety, performance, efficiency, and sustainability through technological, analytical, and regulatory strategies.

The safe operation involves the range of conditions (velocity, temperature, pressure) under which a propulsion system can function reliably and safely. Improving

IntechOpen

the safe operation includes advanced materials and coatings, active control systems, and integrated monitoring systems with onboard sensors. To reduce environmental impact (e.g., to minimize greenhouse gas emissions and noise pollution), sustainable fuels, and combustion techniques, schemes with improved aerodynamic and thermodynamic efficiency are designed and applied [3].

3. Computational and experimental tools

The gas dynamics of internal flows, thermodynamics, and combustion processes are fundamental to the design, performance, and efficiency of propulsion systems, from jet engines to rockets and internal combustion engines. These areas are deeply interconnected and govern how propulsion systems convert chemical energy into thrust or mechanical work [4].

Computational techniques include computational fluid dynamics (CFD) used for analyzing internal flows in turbines, compressors, combustion chambers, and finite element method (FEM) applied to assess stresses, and thermal and vibration responses of mechanical components. CFD solves the Navier–Stokes equations numerically and closing equations and relations to simulate fluid flow, heat transfer, and chemical reactions, allowing engineers to optimize combustion and aerodynamics before building physical prototypes. FEA is used to analyze structural integrity, thermal stresses, vibration, and fatigue life of engine components under real operating conditions. CFD predicts pollutant formation zones and allows redesign of combustion chambers and injectors to reduce emissions. FEA ensures high-temperature durability of exhaust systems and catalytic converters.

The demand to understand the fundamental physics of fluid flows, heat and mass transfer, and combustion mechanisms is driven by the need to design safer, more efficient and environmentally sustainable propulsion and energy systems [5]. This foundational knowledge underpins innovation in virtually all types of propulsion (gas turbines, rocket engines, internal combustion engines, and electric propulsion systems). These principles are critical to predict aerodynamic of fluid flow over blades and airfoils and internal flows in the ducts, to reduce drag and increase thrust or efficiency and combustion (control of fuel/air mixing and flame dynamics leads to more complete combustion, improving energy output per unit of fuel) efficiency, reduce pollutant emissions (NOx, CO_2, and unburned hydrocarbons emissions). Combustion instabilities can cause violent pressure oscillations (e.g., in rocket engines or afterburners). Flow separation or stall in compressors can lead to engine failure. Predicting and controlling these relies on deep fluid dynamics knowledge. Better turbulence and flow control means more homogeneous combustion, reducing hotspots that cause pollutant formation.

Improving the performance of propulsion and power combustion systems requires a detailed understanding of multiphase turbulent reacting flows [6]. Proper atomization, droplet breakup, and evaporation in sprays are critical for efficient combustion. Turbulent eddies affect how well the fuel and air mix, for example, poor mixing leads to incomplete combustion, high emissions, and hotspots. Turbulence affects ignition delay, flame propagation speed, and flame anchoring. Multiphase interactions (e.g., uneven droplet evaporation) can lead to misfires or unstable flames in gas turbines and engines. The rate of heat release couples with the flow field causing instabilities in propulsion systems like combustion oscillations.

4. High-temperatures effects

High-temperature gas effects are also important in high-speed aerodynamic design and optimization, especially when dealing with supersonic and hypersonic vehicles. At these extreme speeds (typically Mach 5 and beyond), the kinetic energy of the vehicle is converted into thermal energy due to the intense friction between the air (or other gases) and the vehicle surface. This results in significant aerodynamic heating, which can have major implications for both the design and safety of high-speed vehicles [7].

The development and implementation of methods and tools that adequately model fundamental physics and allow for credible physics-based optimization is crucial for ensuring the flight safety of high-speed vehicle systems. As vehicle speeds increase, especially in the supersonic and hypersonic regimes, the complexity of physics-based modeling increases dramatically. Optimizing these systems for safety, efficiency, and performance requires detailed and accurate simulations to predict and prevent issues before they arise in real-world flight conditions.

The aero-thermodynamic design and optimization of high-speed vehicles involves a multi-faceted approach, combining wind tunnel testing, flight experiments, and computer simulations. Each of these methods plays a critical role in ensuring that the vehicle performs efficiently and safely under the extreme conditions encountered at supersonic and hypersonic speeds. Wind tunnel testing remains a vital tool for validating aerodynamic and thermodynamic designs. At high speeds, physical models of the vehicle are tested in subsonic, transonic, supersonic, or hypersonic wind tunnels to gather real-world data about aerodynamic forces, heat loads, and flow characteristics. While wind tunnel testing and computer simulations are incredibly useful, actual flight experiments provide the most accurate data for understanding the performance of high-speed vehicles in real-world conditions. These tests provide flight data that cannot always be replicated in controlled environments. Flight experiments give engineers insights into real aerodynamic behavior, including unsteady aerodynamic forces, control surface effectiveness, and maneuverability. Flight tests provide the opportunity to measure aerodynamic heating at actual operational conditions, allowing engineers to assess the effectiveness of thermal protection systems and cooling strategies.

The uncertainty in aerodynamic and propulsion characteristics for high-speed vehicles is a significant challenge. Despite advances in computer simulations and wind tunnel testing, there are still several areas where accurate prediction of behavior remains elusive. This uncertainty is often due to a combination of factors like complex flow dynamics, extreme flight conditions, and limitations in experimental data.

5. Conclusion

Understanding fluid flow, heat transfer, and combustion physics is essential for safe operation (avoidance of stall, suppress of detonation), performance gains (thrust, efficiency), emission compliance (meeting environmental regulations), and innovation (in hypersonic, hybrid-electric, and green propulsion). Localized rich/lean pockets in a turbulent flame zone directly influence formation of NOx, CO, soot, and unburned hydrocarbons. Accurate modeling of these zones requires resolving both fluid turbulence and chemical kinetics.

The importance of propulsion and energy systems research lies in their central role in enabling motion, power generation, and technological advancement across a wide range of industries. Computational modeling tools are a virtual experimental bench for studying the features of the technological process of obtaining products, calculating modes and parameters of processes that ensure the synthesis of new materials and smart products.

Author details

Konstantin Volkov
Kingston University, London, United Kingdom

*Address all correspondence to: k.volkov@kingston.ac.uk

References

[1] Smirnov NN. Space flight safety: Past and future. Acta Astronautica. 2022;**194**:353-357

[2] Smirnov NN. Fundamental problems of safety in space flights. Acta Astronautica. 2024;**219**:871-874

[3] Gohardani AS, Stanojev J, Demaire A, Anflo K, Persson M, Wingborg N, et al. Green space propulsion: Opportunities and prospects. Progress in Aerospace Sciences. 2014;**71**:128-149

[4] Bulat PV, Musteikis AI, Prodan NV, Renev ME, Volkov KN. Simulation of intra-chamber processes in a low-thrust rocket engine with a hydrogen-air mixture and counterflow cooling. Acta Astronautica. 2024;**225**:243-251

[5] Emelyanov VN, Volkov KN, Yakovchuk MS. Outflow of a supersonic overexpanded air jet into a water. European Journal of Mechanics B/Fluids. 2024;**105**:1-13

[6] Volkov K. Flow and aeroacoustic characteristics of underexpanded supersonic jets exhausting from a conical converging nozzle. Fluids. 2024;**9**:148

[7] Smirnov NN. Space flight safety – Discussing perspectives. Acta Astronautica. 2016;**126**:497-499

Chapter 2

Multiscale Modeling and Simulation on Active Cooling Process of Supercritical-Pressure Hydrocarbon Fuel in Ramjet Engines

Xing Sun, Jialiang Wang, Tingting Jing, Fei Qin and Yuepeng Xin

Abstract

As the flight speed increases to the hypersonic regime, thermal management becomes a significant challenge. An effective and reliable thermal protection system (TPS) plays an important role in ensuring the operability of ramjet engines. Regenerative cooling technology is widely adopted in current TPS. During the cooling process, a complex physicochemical process of hydrocarbon fuels at supercritical pressure in the regenerative cooling channel will occur, exerting significant impacts on cooling efficiency. This chapter will introduce the multiscale modeling and simulation of the regenerative cooling process of supercritical-pressure hydrocarbon fuels in ramjet engines. Firstly, a comprehensive and systematic review of the current research status on the thermal structure of advanced propulsion systems applying superalloy or carbon fiber-reinforced ceramic composites will be given. Then, the numerical modeling and simulation techniques describing the flow, heat transfer, thermal oxidation coking, and pyrolytic coking phenomena of hydrocarbon fuels at supercritical pressures will be discussed. Finally, multiscale numerical modeling and simulation techniques for thermal-mechanical analysis of composite materials were given.

Keywords: advanced propulsion systems, composites, thermal protection, hydrocarbon fuels, numerical simulation

1. Introduction

Combined-cycle engines (CCE), such as the Rocket-Based-Combined-Cycle (RBCC) and Turbine-Based-Combined-Cycle (TBCC) propulsions, are expected to be the most efficient propulsion systems for the advanced reusable launch vehicles (RLVs) [1–3]. As the flight Mach number increases, thermal management in the ramjet combustor of the CCE becomes a significant challenge. When the flight Mach number exceeds 8, the total gas temperature in the combustion chamber approaches 3000 K, surpassing the temperature limits of existing metal materials, including superalloys and composites. Thus, thermal protection has become a key factor restricting the development of CCE [4, 5].

2. Thermal protection method

Thermal protection technology has experienced the development of metal heat sink passive thermal protection, thermal barrier coating thermal protection, composite passive thermal protection, heat-absorbing hydrocarbon fuel regenerative cooling, and more advanced combined thermal protection techniques. Ramjet engine thermal protection mainly includes active thermal protection and passive thermal protection. Active thermal protection is the use of "low-temperature" fluid (compared to the hot gas) for engine cooling, while passive thermal protection is the use of material with good heat-resistant properties (such as high-temperature ablation resistance, high heat sink, and low thermal conductivity, high radiation) to directly resist the high heat flux. The definition, merits, and demerits of these methods are listed in **Table 1**. A more detailed discussion of these methods is conducted in the following section.

Type	Categorization	Principle	Specificities
Passive cooling	Ablation cooling	Carrying away heat through material mass consumption.	Advantages: simple structure, high technological maturity, low structural weight. Disadvantages: Limited by strength, temperature limit, ablative deformation, and other factors, cannot meet the engine's long-term reusable thermal protection needs.
	Thermal barrier coating cooling	Ceramic materials cover the surface of metal alloys to improve resistance to high-temperature oxidation.	
Active cooling	Film cooling	Incoming air forms a film between the hot air stream and the wall surface, impeding heat transfer.	Advantages: Lightweight structure, good cooling effect when the temperature of incoming air is low. Disadvantages: Complicated structure process, with the incoming air temperature increasing, the air film cooling effect is reduced.
	Regenerative cooling	Uses fuel as a coolant that flows in cooling channels within the walls of the combustion chamber to absorb heat.	Advantages: Achieve steady-state thermal equilibrium so that the structure can work for a long time, utilize the heat to preheat the fuel in advance, and improve combustion efficiency. Disadvantage: Hydrocarbon fuels have limited heat sinks and can suffer from coking and carbon precipitation in the cooling channels, leading to cooling failure.
	Transpiration cooling	The coolant, under applied pressure, seeps into the combustion chamber and forms a sweating stream within the adhering surface layer.	Advantages: Lower coolant demand, high cooling efficiency, less fuel waste in the near-wall zone compared to liquid film cooling. Disadvantage: Problems with large engine applications due to transpiration patterns, flow distribution, etc.

Table 1.
Comparison of active and passive cooling methods.

2.1 Passive thermal protection

2.1.1 Insulation ablation cooling

Chemical reactions will occur to ablation-resistant materials in high-temperature and high-heat flow conditions; the material will undergo melt, evaporation, decomposition, carbonization, or combustion through the mass consumption on the material surface to take away most of the heat. At the same time, the reaction of the gas generated in the vicinity of the material forms a pneumatic thermal resistance to prevent the heat transfer to the internal structure, thus the internal structure is protected.

2.1.2 Thermal barrier coating cooling

Thermal barrier coating is a thermal protection technology that covers high-temperature, highly insulating ceramic materials on the surface of metal alloys. This improves the base alloy's ability to resist high-temperature oxidation and thus reduces the alloy's surface operating temperature.

2.2 Active thermal protection

2.2.1 Regenerative cooling

The use of on-board fuel as a cooling source for direct or indirect cooling effectively utilizes the redundant energy of the fuel and avoids the additional loading associated with the use of other coolants, as shown in **Figure 1a**. When the fuel absorbs heat and reaches temperatures above 750 K, it also undergoes a cracking and heat absorption reaction, which can increase the heat sink of the fuel and enhance the cooling performance of the system through changes in chemical energy. The resulting regenerative cooling technology is an effective means of thermal protection for advanced aerospace engines.

2.2.2 Film cooling

Air film cooling mainly utilizes the incoming air to cool the combustion chamber, as shown in **Figure 1b**. The amount of air in the ram engine can be divided into two parts: One part is into the combustion chamber and fuel mixing combustion, and the other part of the flow is into the cold air channel, in the vicinity of the wall along the tangential direction or with a certain angle of incidence injection. This cold air is in the mainstream pressure and friction under the action of the downstream bending, adhesion in the vicinity of the wall, and the formation of a cold air film with a relatively low temperature.

2.2.3 Transpiration cooling

For hypersonic ramjet engines with higher Mach numbers ($Ma > 10$), transpiration cooling is an effective means of thermal protection. The principle is that the coolant is in the applied pressure, with a small amount from the heated wall outflow into the combustion chamber, and the cooling medium in the attached surface layer forms a transpiration flow, the formation of a uniform and continuous membrane structure in

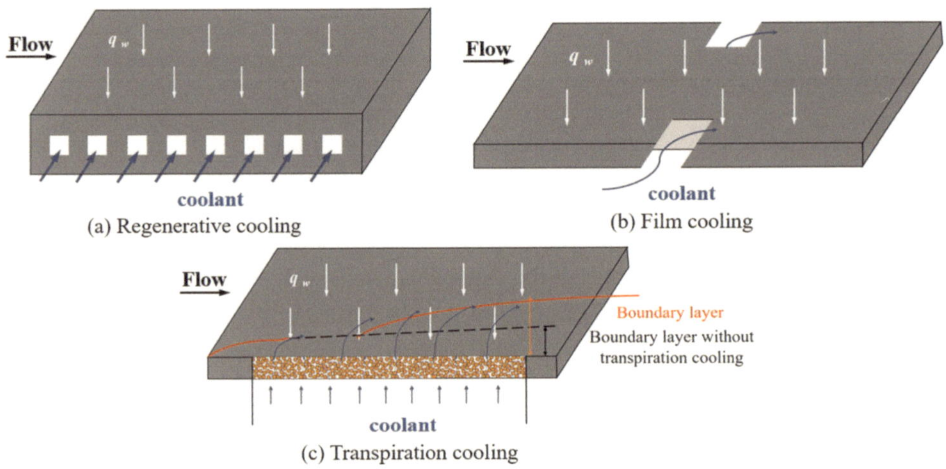

Figure 1.
Main forms of active cooling. Copyright (2020), with permission from Publisher.

the protected wall surface, changing the wall surface attached surface layer of the flow and heat transfer characteristics, significantly reducing the gas through the attached surface layer of the wall of the heat transfer. On the other hand, the cooling medium itself is in the vicinity of the wall surface of the absorption of heat to exert the wall cooling effect, as shown in **Figure 1c**.

2.2.4 Regenerative/film composite cooling

Using partially cracked or uncracked high-temperature fuel (700–1000 K) at the outlet of the regenerative cooling channel as the coolant for air film cooling, the regenerative/air film composite cooling method using hydrocarbon fuel as the coolant is used to cool down localized high-temperature locations on the wall without increasing the mass flow rate of the additional fuel, as shown in **Figure 2**.

Among the above cooling methods, since the increase of flight Mach number has far exceeded the increase of material properties, passive thermal protection technology is difficult to adapt to the harsh thermal field environment caused by aerodynamic heating and combustion exotherm of the aircraft under higher Mach number, and safe and reliable active cooling is the best way to solve the lack of passive cooling capacity. Regenerative cooling has many advantages over traditional cooling methods:

1. Regenerative cooling uses fuel as the cooling medium, which avoids the problem of traditional aircraft needing to carry a large amount of additional cooling medium, reduces the load of the aircraft to a certain extent, conforms to the lightweight design goal of the aircraft, and indirectly optimizes the performance of the aircraft.

2. Regenerative cooling recycles the heat released by combustion, preventing engine components from failing due to high temperatures and increasing the efficiency of the fuel when it subsequently takes part in a reaction.

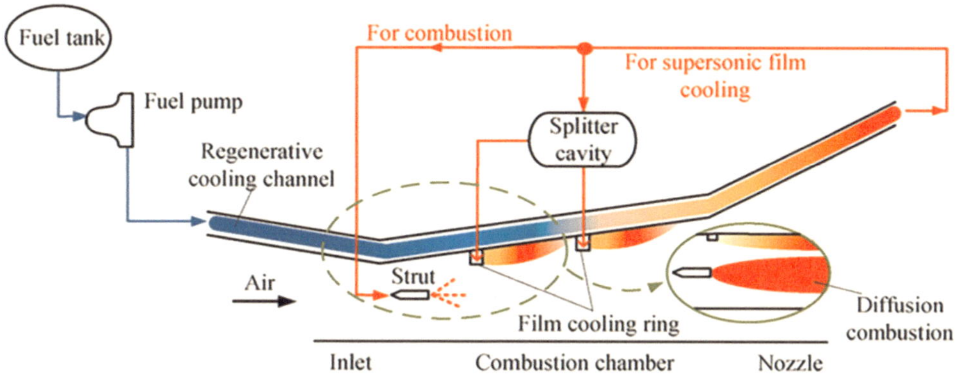

Figure 2.
Regenerative/film composite cooling method [6]. Copyright (2018), with permission from Publisher.

Currently, regenerative cooling is considered to be the best cooling method for super-combustion ramjet engines, and it has been a hot research topic in recent years around the world.

3. Physicochemical process in the regenerative cooling channel

When the pressure in the regenerative cooling channel exceeds the critical pressure of the fuel, the fuel is in a supercritical pressure state. In this state, a fluid with a temperature below its critical temperature is called a compression fluid, while a fluid with a temperature above its corresponding critical temperature is called a supercritical fluid. The transition of a fluid between the two states is affected by temperature and pressure, as shown in **Figure 3**. Below the critical pressure, substances appear as a single phase, two phases (phase equilibrium curve), or three phases coexisting (three-phase equilibrium point), with different substances having different phase equilibrium curves. The transformation process between phases is called the phase transition process, which is often accompanied by energy conversion. The concept of phase transition heat is based on this property of matter, but the discontinuous change of fluid physical properties during the phase transition heat process is a great challenge for theoretical research and practical application. However, when the temperature or pressure exceeds the critical value of the fluid and the fluid is transformed from gaseous state or liquid state to supercritical state, the change of physical properties is a continuous process, and there is no obvious phase transition interface, which is one of the important reasons for choosing supercritical fluid as the cooling medium.

3.1 Physical changes in supercritical hydrocarbon fuels

The physical properties of supercritical fluids are characterized by the high density of liquids and the low viscosity and high diffusivity of gases, and they are widely used in various industrial facilities, such as aerospace, nuclear energy, refrigeration, and so on. **Table 2** lists the critical temperatures and critical pressures of the masses that are widely used in the above fields.

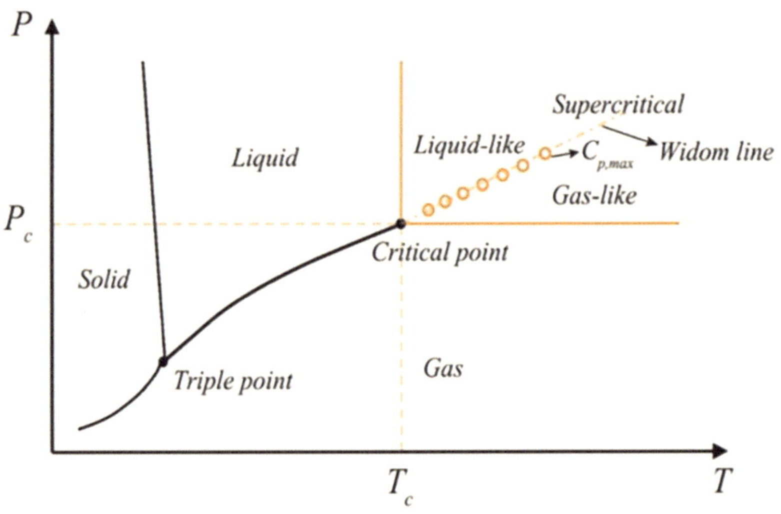

Figure 3.
P–T phase diagram for a substance.

Substances	Decane	RP-3	Methane	JP-10	CO_2	Water
T_c (K)	617.7	646	190.56	617.55	304.12	647.096
P_c (MPa)	2.1	2.3	4.599	2.097	7.374	22.064

Table 2.
Critical parameters for some substances.

Figure 4 gives the variation of thermodynamic and transport properties of n-decane with temperature at 3 MPa. It can be seen that when the temperature exceeds the critical temperature, the surface tension gradually disappears as the fluid transforms from the liquid state to the supercritical state, and this process leads to a drastic decrease in density, dynamic viscosity, and thermal conductivity, while the constant-pressure specific heat will appear a very high peak, and the temperature corresponding to this peak is often defined as the proposed critical temperature, and the curves of the proposed critical temperatures of the substances at different pressures are called the Widom line [7–10]. The Widom line, shown in **Figure 3**, characterizes the transition of a supercritical fluid from a liquid-like state to a gas-like state, analogous to the phenomenon of boiling at subcritical pressures. As the pressure increases, the intensity of the physical change decreases.

3.2 Flow and heat transfer characteristics of supercritical hydrocarbon fuels

Research on supercritical fluid can be traced back to the 1930s, in the study of turbine blade cooling, Schmidt et al. discovered the special properties of fluid flow heat transfer under supercritical pressure and took the lead in opening the supercritical pressure heat transfer research.

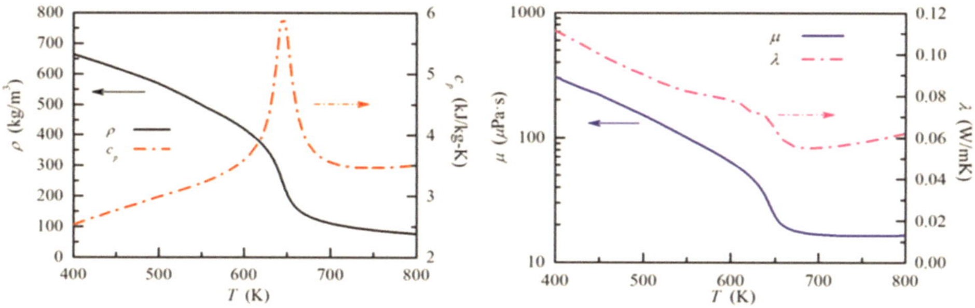

(a)Density and specific heat at constant pressure

(b) Dynamic viscosity and thermal conductivity

Figure 4.
Thermophysical and transport properties of decane at 3 MPa.

Generally, we can classify the flow heat transfer phenomena of supercritical fluids into the following three categories [11, 12].

The first category is normal heat transfer (NHT), which is often characterized by the Dittus-Boelter correlation.

$$Nu = 0.023\,\mathrm{Re}_b{}^{0.8}\mathrm{Pr}_b{}^{0.4} \tag{1}$$

The second category is deteriorated heat transfer (DHT), which is defined by Pioro and Duffey [13] as having higher wall temperatures and lower heat transfer coefficients compared to normal heat transfer. There are a number of criteria to characterize the phenomenon of deteriorated heat transfer, which can be divided into two main categories: those judged by the distribution of wall temperatures and those judged by the heat transfer coefficient or Nusselt number.

Urbano and Nasuti [14] proposed a quantitative criterion for heat transfer deterioration based on the wall temperature derivatives, where a heat transfer deterioration condition is considered when there is a peak in the wall temperature for $(\mathrm{d}T_w/\mathrm{d}x)_{\min}$ or $(\mathrm{d}T_w/\mathrm{d}h_b)_{\min} < 0$. On the contrary, a non-deteriorating condition is considered when there is a monotonic increase in the wall temperature. Another type of evaluation criterion is based on the empirical formula of the Nusselt number or convective heat transfer coefficient. Gupta et al. [15] proposed a modified Dittus-Boelter empirical formula to calculate the Nusselt number for normal heat transfer Nu_{ref}:

$$Nu_{ref} = 0.009\,\mathrm{Re}_b{}^{0.89}\overline{\mathrm{Pr}}^{-0.14}\left(\frac{\rho_w}{\rho_b}\right)^{0.93}\left(\frac{\lambda_w}{\lambda_b}\right)^{0.22}\left(\frac{\mu_w}{\mu_b}\right)^{-1.13} \tag{2}$$

Where $\overline{\mathrm{Pr}} = \mu_b\bar{c}_p/\lambda_b$ and $\bar{c}_p = (h_w - h_b)/(T_w - T_b)$. When $Nu/Nu_{ref} < 1.0$, it was concluded that heat transfer deterioration occurred in this condition. Zhang et al. [11] also proposed a criterion for evaluating heat transfer deterioration based on the convective heat transfer coefficient, when $h/h_{ref} < 1.0$, heat transfer deterioration is thought to occur, among other things:

$$h_{ref} = 0.9 \cdot \left(0.023\,\mathrm{Re}_b{}^{0.8}\mathrm{Pr}_b{}^{0.4}\right) \cdot \frac{\lambda_b}{d} \tag{3}$$

The third type of heat transfer phenomenon is enhanced heat transfer (EHT), this type of phenomenon tends to appear together with the heat transfer deterioration phenomenon, when the heat transfer coefficient exceeds the normal heat transfer, it can be assumed that the working condition of heat transfer enhancement phenomenon.

In the process of supercritical pressure fluid flow heat transfer, the phenomenon of heat transfer deterioration or intensification is relatively common, but the causes and influencing factors are very complicated. The academic community has a more unified view that the heat transfer deterioration and strengthening phenomenon are mainly caused by the buoyancy effect, thermal acceleration effect, and the proposed critical temperature near the intense physical changes in heat (such as pseudo-boiling).

3.3 Endothermic chemical reactions of supercritical hydrocarbon fuels

During supercritical pressure heat transfer of hydrocarbon fuels, when the temperature exceeds 800 K, the fuel undergoes a cracking and heat absorption reaction (cracking is more intense above 900 K), which converts the applied heat energy into the chemical energy of the fuel, thus lowering the temperature of the fuel and enhancing the heat transfer [16, 17].

Ward et al. [18] conducted a detailed experimental study of the endothermic cracking processes of n-decane and n-dodecane. A one-step cracking reaction mechanism containing 18 components was proposed, namely the proportional product distribution (PPD) model, as shown in **Table 3**, which is now widely used in numerical calculations for n-decane with a conversion rate below 25%.

The reaction rate of n-decane is expressed as:

$$\dot{\omega}_{decane} = -k_A [C_{10}H_{22}] \tag{4}$$

The reaction rate constants k_A were calculated using the Arrhenius formula:

$$k_A = k_0 \exp\left(-\frac{E_a}{R_u T}\right) \tag{5}$$

The pre-exponential constant, k_0 and activation energy, E_a are both provided in **Table 3**. R_u is the universal gas constant (8.314 kJ·kmol^{-1} K^{-1}).

3.4 Oxidative coking and pyrolytic coking of supercritical hydrocarbon fuels

For the coking process of heat-absorbing hydrocarbon fuels, the present study concluded that the reaction mechanism is very complex and accompanied by a variety of chemical reactions and physical processes. Coking is divided into two categories: oxidative coking and cracking coking, and cracking coking can be divided into catalytic cracking coking and high-temperature thermal cracking coking. The thermal cracking coking reaction of kerosene is the main source of carbon accumulation in

Chemical reaction	E_a, kJ·mol^{-1}	k_0, s^{-1}
$C_{10}H_{22} \rightarrow 0.153CH_4 + 0.222C_2H_4 + 0.138C_2H_6 + 0.2C_3H_6 + 0.185C_3H_8 + 0.171C_4H_8 + 0.118C_4H_{10} + 0.149C_5H_{10} + 0.137C_5H_{12} + 0.17C_6H_{12} + 0.106C_6H_{14} + 0.147C_7H_{14} + 0.091C_7H_{16} + 0.132C_8H_{16} + 0.04C_8H_{18} + 0.046C_9H_{18} + 0.031C_9H_{20}$	263.7	1.6×10^{15}

Table 3.
One-step proportional product distribution (PPD) model for thermal cracking of decane.

$k_{c0}\left[\mu g \cdot cm^2 \cdot \min^{-1}/(mol \cdot L^{-1})^i\right]$	E_{a1}[kJ/mol]	n	γ
8.68×10^{11}	192.88	0.46	0.074

Table 4.
Kinetic parameters in the surface coking model.

high-temperature pipelines. Kerosene coking inhibition technology has also been in exploration, from the mechanism of analysis, the use of additives in kerosene is an effective way to stop the production of carbon deposits, while the use of inert coatings for pipelines and catalytic coatings have a certain effect.

The simplified cracking coking mechanism proposed by Liu et al. [19] can be used in the numerical simulation to consider the wall coking phenomenon after the cracking reaction. The expression of the MS-II coking model is given by:

$$\frac{dSC}{dt} = k_{c0} \exp\left(-\frac{E_{a1}}{R_u T_w}\right) \exp(-\gamma t) {C_{p1}}^n \tag{6}$$

C_p is the molar concentration of propylene, and other kinetic parameters of the reaction are shown in **Table 4** [20].

4. Modeling and simulation methods for the regenerative cooling process

Turbulence is a nonlinear fluid motion that is spatially irregular and temporally disordered, which exhibits a very complex flow state and is a well-known problem in fluid mechanics. Its complexity is mainly manifested in the stochastic, cyclonic, and statistical nature of turbulent flow. The traditional computational fluid dynamics to describe the turbulence is based on the Navier–Stokes (N-S) equation. According to different degrees of treatment of the N-S equation on the turbulence, numerical simulation methods are mainly divided into three kinds: Reynolds averaging method (RANS), large eddy simulation (LES), and direct numerical simulation (DNS). Direct numerical simulation can obtain accurate information about the turbulence field, which is an effective means to study the turbulence mechanism, but the existing computational resources are often difficult to meet the needs of the high Reynolds number flow simulation, thus limiting the scope of its application. The Reynolds averaging method can calculate the complex flow with a high Reynolds number, but it gives the average motion results, which cannot reflect the detailed information of the flow field turbulence. Large eddy simulation is based on the turbulent kinetic energy transfer mechanism and directly calculates the motion of large-scale eddies, and the influence rules of small-scale eddy motion on large-scale eddies are reflected through the establishment of the subgrid model, which not only can get more dynamic information, such as large-scale eddy structure and properties than Reynolds averaging method, but also saves computational amount than direct numerical simulation, and has been more and more widely developed and applied.

4.1 Reynolds-averaged Navier: Stokes (RANS)

For the horizontal and vertical downward flow heat transfer processes of supercritical pressure hydrocarbon fuels, a large number of numerical simulation studies have shown that the current Reynolds averaged model (RANS) is able to accurately predict the turbulent flow and heat transfer phenomena.

In RANS, the mass, momentum, energy, and component conservation equations are solved in the fluid domain as follows after Favre averaging because the flow process involves more drastic density changes:

$$\frac{\partial(\bar{\rho}\tilde{u}_i)}{\partial x_i} = 0 \tag{7}$$

$$\frac{\partial(\bar{\rho}\tilde{u}_i\tilde{u}_j)}{\partial x_j} = \bar{\rho}g_i - \frac{\partial\bar{p}}{\partial x_i} + \frac{\partial\tilde{\tau}_{eff}}{\partial x_i} \tag{8}$$

$$\frac{\partial(\bar{\rho}\tilde{u}_j\tilde{e}_t)}{\partial x_j} = \frac{\partial}{\partial x_j}\left(\lambda_{eff}\frac{\partial\tilde{T}}{\partial x_j}\right) - \frac{\partial(\bar{p}\tilde{u}_j)}{\partial x_j} \tag{9}$$

$$\frac{\partial(\bar{\rho}\tilde{u}_j\tilde{Y}_k)}{\partial x_j} = -\frac{\partial}{\partial x_j}(\bar{\rho}\tilde{u}_{d,k}\tilde{Y}_k) + \tilde{S}_k \tag{10}$$

In Eqs. (7)–(9), ρ is density, u is velocity, $p$is pressure, g_i is gravity vector, e_t is total energy, T is temperature. For any physical quantity $\varphi(x_i, t)$, "$\sim$" is the density-weighted average or Favre average $\left(\tilde{\varphi}(x_i,t) = \overline{\rho\varphi(x_i,t)}/\bar{\rho}\right)$, for quantities related to velocity, temperature, and internal energy; "-" is the time average, used for averaging density and pressure. The above equations are applied with steady-state calculations and the transient terms are not given.

In the momentum equation, $\tilde{\tau}_{eff}$ is the viscous stress term with the expression:

$$\tilde{\tau}_{eff} = 2\mu_{eff}\left(\tilde{S}_{ij} - \frac{1}{3}\frac{\partial\tilde{u}_l}{\partial x_l}\delta_{ij}\right) \tag{11}$$

μ_{eff} is the sum of the kinetic viscosity and turbulent viscosity $\left(\mu_{eff} = \mu_t + \mu\right)$. $\tilde{S}_{ij}$ is the strain rate tensor, denoted:

$$\tilde{S}_{ij} = \frac{1}{2}\left(\frac{\partial\tilde{u}_i}{\partial x_j} + \frac{\partial\tilde{u}_j}{\partial x_i}\right) \tag{12}$$

Similarly, in the diffusion term of the energy equation, λ_{eff} is the effective thermal conductivity after taking into account turbulent diffusion, which is usually simplified using the simple gradient diffusion model:

$$\lambda_{eff} = \lambda + \frac{\mu_t c_p}{\mathrm{Pr}_t} \tag{13}$$

Pr_t is the turbulent Prandtl number, and previous studies have shown that a value of 0.85 can be applied to current supercritical pressure heat transfer calculations for hydrocarbon fuels.

The component transport equation, $\tilde{Y}_k$ is the mass fraction of component k and $\tilde{u}_{d,k}$ is the diffusion coefficient of component k, defined as:

$$\tilde{u}_{d,k} = -\frac{D_{k,m}}{\tilde{Y}_k}\frac{\partial\tilde{Y}_k}{\partial x_i} \tag{14}$$

$D_{k,m}$ is the component diffusion coefficient for component k.

In cases where there is a chemical reaction, such as a high-temperature cracking reaction of a fuel, the following chemical reaction source term $\tilde{S}_k$ needs to be considered in the component transport equation:

$$\tilde{S}_k = M_{wk}\dot{\omega}_k \tag{15}$$

M_{wk} is the molar mass of component k, and $\dot{\omega}_k$ is the net production rate of the component.

In the part of the calculation involving the flow/solid coupling, the solid domain solves the following thermal conductivity equation:

$$\frac{\partial}{\partial x_j}\left(\lambda\frac{\partial\tilde{T}}{\partial x_j}\right) + \dot{q}^{'''} = 0 \tag{16}$$

4.2 Large eddy simulation

Due to the oversimplification of the RANS model, for example, the simplified turbulent kinetic energy generation term is unable to consider the anisotropic turbulence, and the use of a simple gradient diffusion model leads to the calculation of the buoyancy term being small and other problems, so that the RANS model for the strong role of buoyancy force of the vertical upward flow of the process of heat transfer prediction error is large. The fundamental reason is the inability to solve the turbulence field. Therefore, the more accurate large eddy simulation (LES) method is suitable for the numerical simulation of phenomena such as the upward flow of supercritical fluids and heat transfer in vertical tubes. The basic idea of large eddy simulation is to directly solve for eddies larger than a certain scale, while eddies smaller than a certain scale are handled by a model, which not only improves the accuracy of the calculation but also avoids the huge amount of calculation brought by direct numerical simulation.

4.2.1 Governing equations for large eddy simulation

The Favre filtered large eddy simulation control equations are as follows:

$$\frac{\partial\overline{\rho}}{\partial t} + \frac{\partial(\overline{\rho}\tilde{u}_i)}{\partial x_i} = 0 \tag{17}$$

$$\frac{\partial(\overline{\rho}\tilde{u}_i)}{\partial t} + \frac{\partial(\overline{\rho}\tilde{u}_i\tilde{u}_j)}{\partial x_j} = \overline{\rho}g_i - \frac{\partial\overline{p}}{\partial x_i} + \frac{\partial\tilde{\sigma}_{ij}}{\partial x_j} - \frac{\partial\tau_{ij}}{\partial x_j} \tag{18}$$

$$\frac{\partial\left(\overline{\rho}\tilde{h}\right)}{\partial t} + \frac{\partial\left(\overline{\rho}\tilde{u}_j\tilde{h}\right)}{\partial t} = \frac{\partial}{\partial x_j}\left[\left(\frac{\lambda}{C_p} + \frac{\mu_{sgs}}{\mathrm{Pr}_{sgs}}\right)\frac{\partial\tilde{h}}{\partial x_j}\right] + \overline{\rho}g_i\tilde{u}_i \tag{19}$$

Unlike the RANS model, the "$\sim$" here represents a density-weighted filter or Favre filter for velocity or enthalpy-related quantities. The Favre filter for any physical quantity $\varphi(x_i, t)$, defined by Equation:

$$\tilde{\varphi}(x_i, t) = \overline{\rho\varphi(x_i, t)}/\overline{\rho} \tag{20}$$

"-" for spatial filtering.

$\overline{\rho}g_i$ and $\overline{\rho}g_i\tilde{u}_i$ in Eqs. (18) and (19) represent the action terms of the buoyancy force in the momentum and energy equations, respectively, and $\tilde{\sigma}_{ij}$ is the tangential stress tensor, denoted as:

$$\tilde{\sigma}_{ij} = \mu\left(\frac{\partial \tilde{u}_i}{\partial x_j} + \frac{\partial \tilde{u}_j}{\partial x_i}\right) - \frac{2}{3}\mu\frac{\partial \tilde{u}_k}{\partial x_k}\delta_{ij} \tag{21}$$

τ_{ij} is the subgrid stress tensor, defined as:

$$\tau_{ij} = \overline{\rho}\widetilde{u_i u_j} - \overline{\rho}\tilde{u}_i\tilde{u}_j \tag{22}$$

This item is often treated using the Boussinesq eddy-viscosity hypothesis:

$$\tau_{ij} - \frac{1}{3}\tau_{kk}\delta_{ij} = -2\mu_{sgs}\left(\tilde{S}_{ij} - \frac{1}{3}\tilde{S}_{kk}\delta_{ij}\right) \tag{23}$$

where the strain rate tensor $\tilde{S}_{ij}$, denoted as:

$$\tilde{S}_{ij} = \frac{1}{2}\left(\frac{\partial \tilde{u}_i}{\partial x_j} + \frac{\partial \tilde{u}_j}{\partial x_i}\right) \tag{24}$$

τ_{kk} can be neglected at low-velocity flow and small pressure drop. The sublattice heat flux in the energy equation is treated by the gradient diffusion model, where the sublattice Prandtl number Pr_{sgs} is taken as 0.85.

4.2.2 Subgrid-scale model

For the supercritical-pressure flow and heat transfer problems, the subgrid viscosity μ_{sgs} in the control equation can be calculated using the Wall-Adapting Local Eddy-Viscosity (WALE) model, which has been widely used in heat transfer calculations because it has been shown to have higher accuracy and computational stability when dealing with near-wall turbulence and turbulence turning conditions.

In the model,

$$\mu_{sgs} = \overline{\rho}{L_s}^2\frac{\left(S_{ij}^d\right)^{3/2}}{\left(\tilde{S}_{ij}\tilde{S}_{ij}\right)^{5/2} + \left(S_{ij}^d S_{ij}^d\right)^{5/4}} \tag{25}$$

where the subgrid mixing length scale L_s is defined:

$$L_s = \min(\kappa y, C_w\Delta) \tag{26}$$

In Eq. (26), κ is the von Kármán constant, 0.4187. y is the distance from the grid center point location to the wall. C_w is 0.325. The other parameters are defined as

$$\Delta = V^{1/3} \tag{27}$$

$$S_{ij}^{d} = \frac{1}{2}\left(\tilde{g}_{ij}^{2} + \tilde{g}_{ji}^{2}\right) - \frac{1}{3}\delta_{ij}\tilde{g}_{kk}^{2} \tag{28}$$

$$\tilde{g}_{ij} = \frac{\partial \tilde{u}_i}{\partial x_j} \tag{29}$$

5. Anisotropic and multiscale modeling of supercritical-pressure flow and heat transfer in composite cooling channels

In recent decades, significant efforts have been devoted to exploring new thermal structural materials for both passive and active thermal protection systems in next-generation reusable launch vehicles (RLV). Future two-stage-to-orbit (TSTO) and single-stage-to-orbit (SSTO) missions demand a higher thrust-to-weight ratio. Metal-based regenerative cooling structures, due to their high density, are unable to meet these requirements. This has placed new demands on combustion chamber materials, such as high strength, low density, low thermal expansion, and high thermal stability. Carbon/carbon (C/C) composites, referring to carbon fiber-reinforced carbon matrix composites, are considered promising materials to meet these requirements [21, 22]. However, a major drawback of C/C composites is their pronounced erosion and oxidation when the temperature exceeds 500°C, which severely limits their applications. To protect the carbon fibers and enhance the oxidation resistance of C/C composites, various methods have been proposed, such as surface coatings and matrix modifications. Research has shown that introducing refractory carbides, nitrides, and borides can significantly improve the ultra-high-temperature performance of C/C composites. For example, hafnium carbide (HfC) and silicon carbide (SiC) are widely used to enhance the ablation resistance and oxidation resistance of C/C composites. Both have high melting points, approximately 3000°C and 2700°C, respectively. Moreover, their oxides (SiO_2 and HfO_2) form at high temperatures, acting as barriers to further oxidation of the carbon [23].

Although carbon fiber-reinforced ceramic matrix composites can withstand relatively high temperatures (with C/SiC composites having a long-term service temperature of around 1900 K), the temperatures inside a typical RBCC engine combustion chamber—approximately 3000 K under Mach 7 flight conditions—still exceed the temperature limits of any composite material. For limited operating periods, passive cooling structures can only be applied in low-temperature regions, such as engine inlets. Moreover, due to the poor mechanical properties of early carbon fiber-reinforced composites—such as low tensile strength, high brittleness, and low thermal conductivity—cooling structures that combine metal channels with composite materials have been commonly used in practical applications [24–26]. However, there is a growing trend in active thermal protection systems to increase the proportion of carbon-reinforced composites, which offer lower density, higher specific strength, and superior temperature resistance. With the advancement of computational mechanics and high-performance computing clusters (HPCC), numerical methods have become increasingly valuable and are now widely used to assess the mechanical properties, thermal performance, and manufacturing processes of composite materials. These numerical approaches can significantly reduce the experimental costs and design cycles associated with composites.

Since the last century, the application of carbon fiber-reinforced composites in active cooling systems for thermal protection in advanced propulsion systems has gained increasing attention in many countries. The trend is shifting from combining passive cooling structures with metal-based active cooling structures to the development of fully integrated composite active cooling systems. It can be said that increasing the proportion of carbon fiber-reinforced composites, which offer lower density, higher specific strength, and superior temperature resistance, has become the development direction of this field. However, several factors still constrain the further advancement of these applications. These factors mainly include the following: (1) issues with thermal matching and high-temperature connections between composites and metals due to differences in thermal expansion coefficients; (2) leakage of coolant caused by the porosity of carbon-reinforced composites; (3) local damage and overall structural failure due to the poor mechanical strength of composites; and (4) difficulties in manufacturing complex composite structures. Currently, research and experiments on regenerative cooling structures made from carbon-reinforced composites are still limited. Much of the existing research is focused on theoretical simulations and mechanical testing of small-scale single tubes or simple structures. In the future, more research will be needed to address these challenges.

With the development of computational mechanics and high-performance computing clusters, numerical methods become more and more important, which have been adopted to evaluate the mechanical properties, thermal properties, and manufacturing process of composites. The multiscale numerical methods can greatly reduce the experimental costs.

5.1 Multiscale thermal-mechanical analysis of composite materials

Carbon fiber-reinforced composites possess complex internal structures. Different carbon fiber preforms architectures and various types of matrix can significantly impact the mechanical properties and thermal transfer performance of the final cured composite material, leading to pronounced anisotropic and heterogeneous behavior. Consequently, developing appropriate numerical models to accurately analyze the mechanical strength, damage evolution, and thermal performance of these composites has become a prominent research topic in recent years [23].

Structural simulation of composite materials typically employs two main approaches: macroscopic equivalent methods and multiscale-simulated methods. The macroscopic equivalent method treats the entire composite structure as a homogeneous medium, ignoring the interactions between the reinforcement and matrix phases. This approach is generally used for the engineering design of large structural components or when local mechanism analysis is not a priority, with a focus on analyzing overall macroscopic performance. Multiscale methods, on the other hand, are effective tools for analyzing damage and failure in composites. These methods account for microscale, mesoscale, and macroscale interactions, emphasizing the relationships between different scales and striving to describe their characteristics as accurately as possible. The goal is to establish a quantitative relationship between the microstructure and macroscopic properties of the composite material. At the microscale, various factors must be considered, such as the lay-up of laminate, porosity in fiber bundles and the matrix, and strength reductions following heat treatment. The composite's performance can then be derived from the properties of the fiber filaments and matrix phase, often using homogenization techniques such as inclusion theory, mixture rules, or direct numerical simulations. At the mesoscale, periodic unit cells (also known as representative volume elements [RVEs]) are created to represent different ply or weave structures based on the periodicity of the composite material.

The macroscopic properties of the composite structure can be obtained through the homogenization of the mesoscale RVE model. There are two main types of multiscale methods: hierarchical multiscale and concurrent multiscale approaches. The hierarchical multiscale method involves performing separate calculations at each scale, with the effective properties at each lower scale being passed up to the higher scale. This method does not involve coupling between scales, making it more efficient. However, establishing the constitutive laws at each scale can be complex, especially for macroscopic scale modeling. The concurrent multiscale method, on the other hand, couples different scales simultaneously during calculations. With this coupling, only the constitutive laws of the microscale components (i.e., carbon fiber filaments and the matrix phase) need to be defined, and these are generally simpler in form. However, the coupling of scales in concurrent multiscale methods leads to significantly higher computational costs. In summary, the main challenge of the hierarchical multiscale method lies in the accuracy of the effective constants, while the primary difficulty of the concurrent multiscale method is its computational expense.

The main idea behind multiscale methods is to replace the original heterogeneous material with a globally homogenized material, under the condition that the strain energy of the two systems is either identical or approximately the same. The key to multiscale methods is addressing the microscale and mesoscale behavior of composite materials. Fiber-reinforced composites typically utilize micromechanical methods, which can be divided into analytical and numerical approaches, such as the finite element method (FEM) [27]. Analytical methods are often employed to study the elastic properties of composites and include techniques such as the self-consistent method, generalized self-consistent method, Mori-Tanaka method, and cellular element model [28–31]. Unlike analytical methods, the FEM discretizes the composite structure into a mesh and uses the variational principle to solve the mechanical response of the RVE model, and then homogenizes the properties to the macroscopic scale. Compared to analytical methods, FEM is better suited for dealing with composites that have a high fiber content or complex structures. A significant body of research based on FEM has been conducted to explore the elastic, viscoelastic, plastic, damage failure, and thermodynamic properties of composite materials.

Due to the complexity, heterogeneity, and anisotropy of composite structures, achieving accurate and reasonable simulations of composite materials has been a topic of widespread interest. This is particularly significant for their application in advanced thermal protection systems. Extensive research has been conducted on finite element modeling and analysis of carbon-ceramic composites, with particular emphasis on mechanical anisotropy and the initiation and evolution of material damage. However, research on their thermal properties has been relatively limited. Developing multiscale methods for complex composite structures can significantly reduce computational costs while allowing for precise localization of critical areas and accurate damage simulation. This approach serves as a powerful tool for structural design and optimization.

5.2 Physicochemical process of hydrocarbon fuels in the composite regenerative cooling channel

Another critical issue to consider is the flow and heat transfer process of hydrocarbon fuels in regenerative cooling channels, which could significantly impact the operability of composite materials in thermal protection systems for hypersonic propulsion. As discussed in the previous section, the heterogeneity and anisotropy of composite materials—such as anisotropic thermal conductivity—can greatly affect the heat transfer efficiency within the regenerative cooling channels. In turn, nonuniform heat load distribution may influence the mechanical properties of the composites and affect their fatigue life. Therefore, a detailed study of the physicochemical processes of hydrocarbon fuels within regenerative cooling channels, particularly in composite cooling structures, is essential.

Unlike other supercritical fluids such as water or carbon dioxide, hydrocarbon fuels undergo significant thermochemical reactions during regenerative cooling. These reactions can be broadly divided into two categories based on the reaction temperature: thermal oxidation and surface coking (423–723 K), endothermic pyrolysis, and surface coking (>750 K). Both thermal oxidation coking and pyrolysis coking are non-catalytic coking reactions, resulting from free radical polymerization and dehydrogenation, with coke typically consisting of particulate matter ranging from 1 to 2 μm in size. The morphologies of these two types of surface coke are illustrated in **Figures 5** and **6**. Compared to pyrolysis reactions, the thermal oxidation coking process is much slower. However, over extended heating periods, coke deposition becomes non-negligible, particularly in reusable vehicles, where fuel oxidation coking can easily occur within cooling channels due to its lower activation temperature.

Temperature not only determines which chemical reactions occur but also plays a crucial role in the reaction rates of these two processes. The thermal oxidation coking process is largely independent of temperature variation, while the pyrolysis process, due to the heat absorbed during the breakdown of large molecular compounds into smaller ones (such as methane, ethylene, propylene, and hydrogen), exhibits significant chemical heat absorption. It is important to note that coke particle deposition on the surface gradually covers the inner walls of the cooling channels, thereby reducing the heat transfer efficiency. Both thermal oxidation coking and pyrolysis coking produce carbon deposits that can clog cooling channels, posing a significant threat to engine safety.

Current research on the flow, heat transfer, oxidation coking, and pyrolysis coking of hydrocarbon fuels under supercritical pressure primarily focuses on metallic regenerative cooling channels. For next-generation RLV, particularly SSTO vehicles, composite cooling channels may present a viable alternative. However, there is a scarcity of research in the open literature regarding the heat transfer and thermochemical processes of hydrocarbon fuels within composite cooling channels. Therefore, it is essential to conduct a comprehensive study on the physicochemical

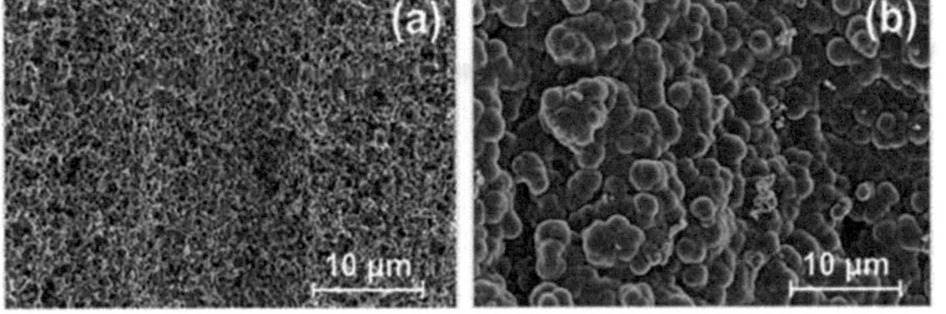

Figure 5.
Morphology of the pyrolytic surface coke deposition, (a) alloy tube without the reaction and (b) JP-10. Reproduced from Ref. [32]. Copyright (2022), with permission from Publisher.

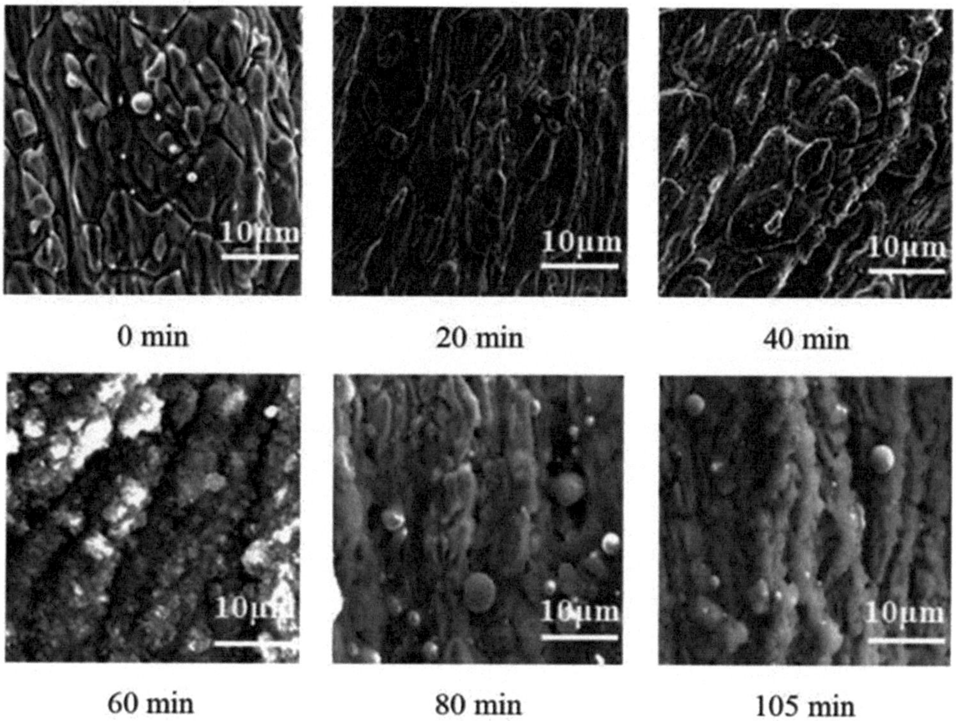

Figure 6.
Morphology of the thermal oxidation coking deposition at different heating conditions. Reproduced from Ref. [33]. Copyright (2018), with permission from Publisher.

processes of hydrocarbon fuels within composite channels. Additionally, when applying composite cooling structures for thermal protection in advanced propulsion systems, it is crucial to consider the coupling effects and interactions between fuel flow and the progressive degradation of the composite materials.

6. Conclusions

Thermal protection plays a dictating role in advanced propulsion systems. This chapter introduces the multiscale modeling and simulation of the regenerative cooling process of supercritical-pressure hydrocarbon fuels in ramjet engines. Firstly, the most commonly used thermal protection tools, including passive and active cooling applying superalloy or carbon fiber-reinforced ceramic composites, were reviewed. It can be concluded that regenerative cooling is the best cooling method for current ramjet engines with metal materials. Then, the physicochemical processes in regenerative cooling channels, including thermophysical property variations, supercritical-pressure flow and heat transfer, endothermic chemical reactions, oxidative coking, etc., were introduced and discussed. The general modeling and simulation methods for these processes were also presented. LES not only can get more dynamic information than the RANS method but also saves computational amount than DNS, which should have more applications in supercritical-pressure flow and heat transfer predictions.

Considering limited coolant supply and superalloy melting point, applying carbon-reinforced composites with lower density, higher specific strength, and higher temperature resistance in active thermal protection structures has become the trend. Multiscale numerical modeling and simulation techniques for thermal-mechanical analysis of composite materials were given. Also, the flow, heat transfer, thermal oxidation coking, and pyrolytic coking phenomena of hydrocarbon fuels at supercritical pressures in the composite cooling channel still lack extensive investigations. Results obtained in this chapter will give practical implications for future hypersonic engine design.

Acknowledgements

The authors appreciate the support of the National Natural Science Foundation of China (Grant No. 52302478), the Natural Science Foundation of Shaanxi (Grant No. 2023-JC-QN-0444), and the National Key Laboratory of Solid Rocket Propulsion (Grant No. SY41YYF202403050).

Conflict of interest

The authors declare no conflict of interest.

Author details

Xing Sun*, Jialiang Wang, Tingting Jing, Fei Qin and Yuepeng Xin
National Key Laboratory of Solid Rocket Propulsion, School of Astronautics, Northwestern Polytechnical University, Xi'an, China

*Address all correspondence to: sunxing@nwpu.edu.cn

References

[1] Czysz P, Little M. Rocket based combined cycle engine (RBCC) - a propulsion system for the 21st century. In: 5th International Aerospace Planes and Hypersonics Technologies Conference. Munich, Germany: American Institute of Aeronautics and Astronautics; 1993

[2] Heiser WH, Pratt DT. Hypersonic airbreathing propulsion. American Institute of Aeronautics and Astronautics. 1994

[3] Sun X, Meng H. Large eddy simulations and analyses of hydrocarbon fuel heat transfer in vertical upward flows at supercritical pressures. International Journal of Heat and Mass Transfer. 2021;**170**:120988

[4] Zhu Y, Peng W, Xu R, Jiang P. Review on active thermal protection and its heat transfer for airbreathing hypersonic vehicles. Chinese Journal of Aeronautic. 2018;**31**:1929-1953

[5] Zhang C, Gao H, Zhao J, Jin H, Wen D. A conjugate numerical model for the thermal analysis of the regenerative cooling of an X-51A-like aircraft. Physics of Fluids. 2023;**35**:097141

[6] Zuo J, Zhang S, Qin J, Bao W, Cui N. Performance evaluation of regenerative cooling/film cooling for hydrocarbon fueled scramjet engine. Acta Astronautica. 2018;**128**:57-68

[7] Simeoni GG, Bryk T, Gorelli FA, Krisch M, Ruocco G, Santoro M, et al. The Widom line as the crossover between liquid-like and gas-like behaviour in supercritical fluids. Nature Physics. 2010;**6**:503-507

[8] Gorelli F, Santoro M, Scopigno T, Krisch M, Ruocco G. Liquidlike behavior of supercritical fluids. Physical Review Letters. 2006;**97**:245702

[9] Sciortino F, Poole PH, Essmann U, Stanley HE. Line of compressibility maxima in the phase diagram of supercooled water. Physical Review E. 1997;**55**:727-737

[10] Banuti DT. Crossing the Widom-line – Supercritical pseudo-boiling. The Journal of Supercritical Fluids. 2015;**98**: 12-16

[11] Zhang Q, Li H, Lei X, Zhang J, Kong X. Study on identification method of heat transfer deterioration of supercritical fluids in vertically heated tubes. International Journal of Heat and Mass Transfer. 2018;**127**:674-686

[12] Zhang S, Xu X, Liu C, Dang C. A review on application and heat transfer enhancement of supercritical CO2 in low-grade heat conversion. Applied Energy. 2020;**269**:114962

[13] Pioro IL, Duffey RB. Experimental heat transfer in supercritical water flowing inside channels (survey). Nuclear Engineering and Design. 2005; **235**:2407-2430

[14] Urbano A, Nasuti F. Conditions for the occurrence of heat transfer deterioration in light hydrocarbons flows. International Journal of Heat and Mass Transfer. 2013;**65**:599-609

[15] Gupta S, Saltanov E, Mokry SJ, Pioro I, Trevani L, McGillivray D. Developing empirical heat-transfer correlations for supercritical CO2 flowing in vertical bare tubes. Nuclear Engineering and Design. 2013;**261**:116-131

[16] Gascoin N, Romagnosi L, Fedioun I, Steelant J, Fau G, Bouchez M. Pyrolysis

in porous media: Part 2. Numerical analysis and comparison to experiments. Journal of Porous Media. 2013;**16**: 857-873

[17] Xu K, Meng H. Modeling and simulation of supercritical-pressure turbulent heat transfer of aviation kerosene with detailed pyrolytic chemical reactions. Energy & Fuels. 2015;**29**:4137-4149

[18] Ward TA, Ervin JS, Striebich RC, Zabarnick S. Simulations of flowing mildly-cracked Normal alkanes incorporating proportional product distributions. Journal of Propulsion and Power. 2004;**20**:394-402

[19] Jiang R, Liu G, Zhang X. Thermal cracking of hydrocarbon aviation fuels in regenerative cooling microchannels. Energy and Fuels. 2013;**27**(5):2563-2577

[20] Xu K, Sun X, Meng H. Conjugate heat transfer, endothermic fuel pyrolysis and surface coking of aviation kerosene in ribbed tube at supercritical pressure. International Journal of Thermal Sciences. 2018;**132**:209-218

[21] Xu LY. Influence of stacking sequence on the transverse matrix cracking in continuous fiber crossply laminates[J]. Journal of Composite Materials. 1995;**29**(10):1337-1358

[22] Yin S, Wang X, Suo X, et al. Deposition behavior of thermally softened copper particles in cold spraying. Acta Materialia. 2013;**61**(14): 5105-5118

[23] Jing T, Xin Y, Zhang L, Sun X, He C, Wang J, et al. Application of carbon fiber-reinforced ceramic composites in active thermal protection of advanced propulsion systems: A review. The Chemical Record. 2023;**23**(4): e202300022

[24] Paquette E, Warburton R, Eckel A, Jaskowiak M. 38th AIAA/ASME/SAE/ ASEE Joint Propulsion Conference and Exhibit. Indianapolis, Indiana: American Institute of Aeronautics and Astronautics; 2002

[25] Paquette E. 13th International Space Planes and Hypersonics Systems and Technologies Conference. Capua, Italy: American Institute of Aeronautics and Astronautics; 2005

[26] Bouchez M, Beyer S. Ptah-socar fuel-cooled composite materials structure. Progress in Propulsion Physics. 2009;**1**: 627-644

[27] Zheng XT, Gou LH, Han SY, et al. Experimental and numerical study on the mode I delamination toughness of z-pinned composite laminates. Key Engineering Materials. 2010; **417**:185-188

[28] Hershey AV. The elasticity of an isotropic aggregate of anisotropic cubic crystals. Applied Mechanics. 1954

[29] Hill R. A self-consistent mechanics of composite materials. Journal of the Mechanics and Physics of Solids. 1965;**13** (4):213-222

[30] Benveniste Y. A new approach to the application of Mori-Tanaka's theory in composite materials. Mechanics of Materials. 1987;**6**(2):147-157

[31] Ghosh S, Lee K, Moorthy S. Multiple scale analysis of heterogeneous elastic structures using homogenization theory and Voronoi cell finite element method. International Journal of Solids and Structures. 1995;**32** (1):27-62

[32] Li H, Wang Y, Wang L, et al. Pyrolysis and coke deposition of JP-10 with decalin in regenerative cooling

DOI: http://dx.doi.org/10.5772/intechopen.1008291

channels. Energy & Fuels. 2022;**36**(12):
6096-6108

[33] Pei X, Hou L, Roberts WL.
Experimental and numerical study on
oxidation deposition properties of
aviation kerosene. Energy & Fuels. 2018;
32(7):7444-7450

Chapter 3

Numerical Analysis of Heat Transfer in Pin-Fin Heat Sinks Using Mini-Channel Configurations

Nabil Bessanane

Abstract

Pin-fins are widely utilized to enhance heat transfer surfaces and induce turbulent flow, thereby improving cooling efficiency through increased heat dissipation, particularly in applications such as hydrogen fuel cells. Over the past decade, their use has become critical in various industrial systems. This study presents a numerical investigation of flow dynamics and heat transfer in rectangular mini-channels (RMC) and pin-fin heatsinks (PFHS), focusing on diamond-shaped pin-fins arranged in a segregated, corrugated configuration. To accurately determine the heat transfer coefficient within this complex thermal system, key parameters—including mass flow rate, geometric dimensions, heat flux, and reference temperature—are thoroughly analyzed. Special emphasis is placed on the method of calculating the reference temperature, as its precise estimation is crucial for optimizing cooling performance. The study introduces a novel technique to accurately determine the temperature difference between the cooling fluid and the heatsink wall, leading to an improved approach for estimating the heat transfer coefficient. A variable reference temperature (VRT) method is employed to calculate the wall-fluid temperature difference. Both qualitative and quantitative analyses are conducted to evaluate flow and heat transfer characteristics. The numerical methodology is validated through experimental measurements, demonstrating that the proposed reference temperature calculation method provides a more accurate representation of the heat transfer coefficient. Additionally, a new correlation is developed to describe the relationship between the averaged heat transfer coefficient and the Reynolds number. This work offers a refined approach to heat transfer coefficient estimation, contributing to the optimization of cooling processes in advanced thermal systems.

Keywords: heat sink, pin-fin, heat transfer coefficient, reference temperature, turbulence, complex geometries, CFD

1. Introduction

Pin-fins are frequently used in cooling devices to enhance heat dissipation by increasing surface area and promoting turbulence. Various types of fins and pin-fins are employed in both natural and forced convection heat transfer applications. The geometric arrangement and dimensions of pin-fins, as well as their thermal-fluid properties, have been extensively studied in the literature, as highlighted by researchers such as Zukauskas and Moores [1, 2] and Shkarah et al. [3].

In addition to conventional designs, the use of pin-fin heat sinks with micro/mini-channels has proven to be highly effective for cooling systems, particularly in applications requiring high heat dissipation within a compact volume, as demonstrated by Obot [4], Kandlikar and Grande [5, 6], and Bahrami et al. [7]. These heat sinks are typically fabricated from materials with high thermal conductivity, such as copper, aluminum, and their alloys, using precision machining or microfabrication techniques. Copper is widely used due to its excellent thermal conductivity, but aluminum and its alloys are also commonly employed, especially in applications where weight and cost are critical factors. The choice of material depends on the specific requirements of the application, including thermal performance, weight, and manufacturability.

The hydraulic diameter (D_h) of these micro-channels typically ranges from 10 to 1000 μm, depending on the coolant fluid passage space. The classification of mini-channels and micro-channels has been widely discussed in the literature [8–11]. However, challenges related to machining and measurement accuracy can lead to errors in key parameters such as hydraulic diameter or reference quantities, complicating the evaluation of heat sink performance, as noted by Morini [12]. One significant challenge is the variation in coolant temperature along the channel, which affects the reference temperature used in heat transfer calculations. These constraints have spurred further research into optimizing heat transfer in cooling systems, particularly for applications like optical cooling.

A key focus of recent research has been the development of methods to accurately calculate the local and average heat transfer coefficients, which are critical for optimizing thermal performance. This involves a detailed evaluation of parameters such as heat flux, wall temperature, and fluid temperature. Numerous numerical simulations have been conducted to assess the performance of thermal systems, with particular attention paid to the effects of reference temperature on heat transfer coefficients. These studies have been validated against experimental results, ensuring their reliability.

Research on flow dynamics in micro-electromechanical systems (MEMS) has revealed phenomena that deviate from those observed in larger-scale systems, complicating the application of conventional thermal-fluid approaches. For instance, Jiang and Xu [13] found that staggered pin-fin arrangements are more efficient than aligned configurations, offering lower pressure drops and better performance at low Reynolds numbers. Diamond-shaped pin-fins, in particular, have been shown to provide superior performance at higher Reynolds numbers.

Despite these advancements, the heat transfer coefficient for complex geometries like pin-fin heat sinks (*PFHS*) remains insufficiently documented in the literature. This is largely due to the challenges associated with accurately evaluating various parameters, especially in small-scale systems where the fluid passage size is comparable to the solid regions, leading to significant temperature variations in the fluid. The reference temperature for the fluid has been defined in various ways, including constant bulk temperature and variable temperature approaches. For example, Liu et al. [14] used the arithmetic mean of inlet and outlet temperatures as the reference temperature in their experimental study of diamond-shaped pin-fins. Yang et al. [15] proposed using the temperature difference ($\Delta T_m = T_w - T_{bulk}$) to evaluate the heat transfer coefficient, where T_{bulk} is the average temperature of the entire fluid domain. However, this approach can lead to unrealistic values if there are small errors in temperature measurement. Mebarki and Rahal [16] suggested that the integral method for calculating bulk temperature is more effective for simple mini-channel heat sinks. Tullius et al. [17] used a "near-wall" temperature as the reference but did not provide detailed methodology. Yang et al. [18] also used the temperature difference ($\Delta T_m = T_w - T_{bulk}$) and achieved good agreement between numerical simulations and analytical results. Eren and Caliskan [19] experimentally analyzed the effect of grooves in pin-fins and used the arithmetic mean of inlet and outlet temperatures as the bulk temperature.

In summary, while significant progress has been made in understanding and optimizing pin-fin heat sinks, challenges remain in accurately determining the heat transfer coefficient, particularly in complex geometries. This study aims to address these challenges by proposing a refined approach to calculating the reference temperature, thereby improving the accuracy of heat transfer coefficient estimation and enhancing the overall performance of cooling systems.

2. Problem description

The mini-channel heat sinks considered in this study are used in the cooling process of the Synchrotron SOLEIL components (monochromators, absorbers, mirrors, etc.). SOLEIL, an acronym for "Optimized Source of Intermediate Energy Light's of LURE (Laboratory for the Use of Electromagnetic Radiation)," is a research center in Paris (France) that operates a particle (electron) accelerator producing synchrotron radiation. This radiation is an extremely powerful source of light that permits the exploration of inert or living matter. Two types of mini-channel heat sinks were studied in this numerical investigation: the heat sink with diamond-shaped pin-fins and the heat sink with simple rectangular mini-channels. The staggered arrangement of pin-fins has been adopted in our study, as it is widely recognized in the literature for its superior heat transfer enhancement capabilities [20–26].

The specific dimensions of the heat sink ($L = 100\ mm$, $W = 16\ mm$, and $H = 1.6\ mm$) were chosen based on the requirements of cooling Synchrotron SOLEIL components, such as monochromators, absorbers, and mirrors. These components operate under high thermal loads, necessitating efficient heat dissipation. The selected dimensions were optimized to ensure:

1. *Effective heat transfer*: The length ($L = 100\ mm$) provides sufficient surface area for heat dissipation, while the height ($H = 1.6\ mm$) and width ($W = 16\ mm$) are designed to accommodate the mini-channel configuration, which enhances heat transfer by promoting turbulent flow and increasing the surface area-to-volume ratio.

2. *Flow characteristics*: The dimensions were chosen to ensure proper flow distribution and minimize pressure drop, which is critical for maintaining efficient cooling performance in the Synchrotron SOLEIL system.

These dimensions were also validated through prior experimental and numerical studies, ensuring their suitability for the specific application. The design aligns with the need for compact, high-performance cooling solutions in advanced scientific equipment like Synchrotron SOLEIL.

The heat sink shape is a parallelepiped (**Figure 1**). The heat sink length is L = 100 mm, and its width and height are W = 16 mm and H = 1.6 mm, respectively. All fins have the same height of 1.6 *mm* and square base dimensions, with the fin width C = 1.7 *mm* (**Figure 2**). The top and bottom of the domain are planes with a thickness of *3 mm*. The distributor length (inlet) is about *5 mm*, which is sufficient for flow establishment. The collector length (outlet) is equal to *12 mm*. The origin plot is fixed at *5 mm*, and the downstream evolution ends at *105 mm*.

This configuration allows for a detailed examination of the thermal and hydrodynamic performance of the heat sinks under various operating conditions, providing insights into their effectiveness in cooling high-power components like those used in the Synchrotron SOLEIL.

In this study, the reference heat dissipater features diamond-shaped pin-fins within mini-channels, maintaining a constant bulk volume. Various geometries with different fluid passage dimensions (e) were examined, though only two specific cases are presented here. These cases were analyzed under varying flow rates ranging from 0.05 to 3 *kg/min*. Additionally, an enhanced near-wall treatment was implemented to account for the highly restricted fluid passage, ensuring a more accurate representation of edge effects on both the thermal and hydrodynamic fields. This approach allows for a detailed investigation of heat transfer and fluid dynamics in such compact and complex geometries.

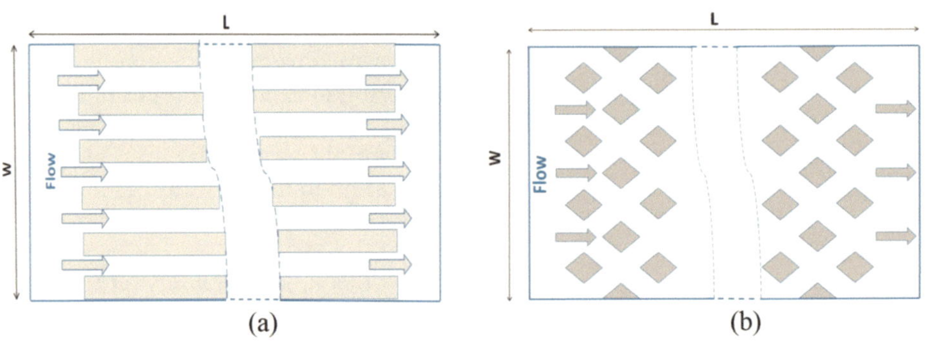

Figure 1.
Schematic presentation of the studied geometries: (a) RMC and (b) PFHS.

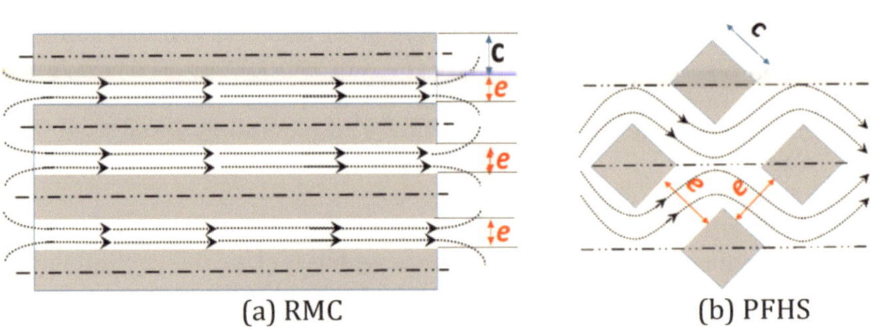

Figure 2.
Fluid passage area for geometries: (a) regular and (b) corrugated.

3. Pin geometry

The pin-fins diamond shape consists of a square section (1.7 × 1.7 mm21.7 × 1.7 mm2) oriented at a 45° angle in the direction of the flow. **Figure 2** shows the fluid passage area for fins geometries: with regular channel (a) and corrugated channel (b). The fins are arranged in a staggered configuration, which is better from the heat transfer point of view, as mentioned in the literature survey. The fluid passage spacing values (e = *0.35, 0.7, 1.2, 1.6 mm*) were chosen based on prior experimental and numerical studies that demonstrated optimal heat transfer and hydrodynamic performance for these dimensions. These studies have shown that these specific spacing values provide a balance between maximizing heat transfer enhancement and minimizing pressure drop, which is critical for efficient cooling performance in mini-channel heat sinks [19–21, 27–29].

The fluid passage spacing values are e = *0.35, 0.7, 1.2, and 1.6 mm*.

4. Numerical approach and mathematical formulation

4.1 Governing equations

The governing equations for the steady-state incompressible flow are given by:

4.1.1 Continuity equation

$$\frac{\partial(\rho U_j)}{\partial x_j} = 0 \tag{1}$$

4.1.2 Momentum and energy equations

$$\frac{\partial(\rho U_i U_j)}{\partial x_j} = -\frac{\partial P}{\partial x_j} + \frac{\partial}{\partial x_j}\left(\mu\left(\frac{\partial U_i}{\partial x_j} + \frac{\partial U_j}{\partial x_i}\right) - \left(\frac{2}{3}\delta_{ij}\mu\frac{\partial U_k}{\partial x_k}\right) - \rho\overline{u_i u_j}\right) \tag{2}$$

$$U_j\frac{\partial(\rho T)}{\partial x_j} = \frac{\partial}{\partial x_j}\left(\frac{\lambda}{C_P}\frac{\partial T}{\partial x_j} - \rho\overline{tu_j}\right), \quad where : \overline{tu_j} = \nu_t\frac{\partial T_i}{\partial x_j} \tag{3}$$

The Reynolds-Averaged Navier-Stokes (RANS) approach was chosen for this study due to its computational efficiency and suitability for engineering applications where detailed turbulence structures are not required. While Direct Numerical Simulation (DNS) and Large Eddy Simulation (LES) offer higher accuracy in resolving turbulent structures, they are computationally expensive and require significantly more resources. For the purpose of this study, which focuses on overall heat transfer and hydrodynamic performance, the RANS approach provides a balance between accuracy and computational cost. This is particularly relevant for engineering applications like the cooling of Synchrotron SOLEIL components, where the primary goal is to optimize heat transfer and flow characteristics rather than resolve fine-scale turbulence details [30–32].

Eq. (2) describes the dynamic flow field, incorporating the Reynolds stress term, which is a key component of the Reynolds-Averaged Navier-Stokes (RANS) equations. This formulation is suitable for all flow conditions examined in this study. Similarly, the convection eq. (3) is expressed for turbulent flow, where temperature and velocity fluctuations are correlated through an additional term introduced in the energy equation. Heat conduction within the solid regions is explicitly accounted for in the model. In this study, radiation effects are neglected, and the Realizable k-ε turbulence model is employed to simulate the turbulent flow.

In this study, the thermal conductivity (λ) is assumed to be constant. This assumption simplifies the numerical model while still providing accurate results for the heat transfer analysis of the mini-channel heat sinks. Although thermal conductivity can vary with temperature in some materials, the constant conductivity assumption is valid for the temperature ranges considered in this study, as the variation in thermal conductivity is negligible within the operating conditions of the Synchrotron SOLEIL cooling system [33].

The *thermophysical* properties of the fluid, such as viscosity (μ) and thermal conductivity (λ), are modeled using polynomial laws to capture their variation with temperature. The polynomial laws are expressed as follows:

$$\mu(T) = \mu_0 + \mu_1 T_1 + \mu_2 T_2 + \mu_3 T_3 \tag{4}$$

$$\lambda\,(T) = \lambda_0 + \lambda_1 T_1 + \lambda_2 T_2 + \lambda_3 T_3 \tag{5}$$

Here, μ_0, μ_1, μ_2, μ_3 and λ_0, λ_1, λ_2, λ_3 are coefficients determined from experimental data or empirical correlations for the specific fluid used in the study. These polynomial laws allow for accurate modeling of the thermophysical fluid properties over the temperature range encountered in the mini-channel heat sinks [33].

4.2 Boundary conditions

The boundary conditions can be summarized as follows in **Figure 3**:

- At the inlet, temperature and velocity are imposed; the inlet temperature is fixed to 303 K, while the inlet velocity varies according to the imposed mass flow rate. The range of mass flow rates used in this study is from 0.05 to 3 *kg/min*, covering both low and high flow regimes to ensure a comprehensive analysis of the heat transfer and hydrodynamic performance of the mini-channel heat sinks.

- At the outlet, all the variable gradients are fixed to the Neumann condition, except for the pressure, which is imposed to 3 bars (the nominal parameter given by experimental data).

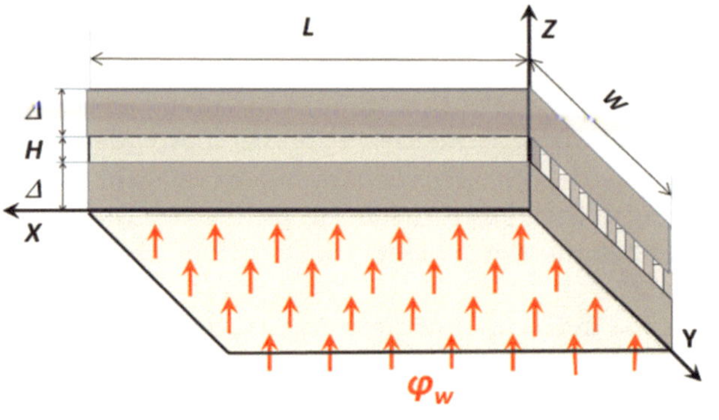

Figure 3.
Geometric characteristics of the heat sink model.

- A no-slip condition is considered for all inside walls, and an adiabatic condition is applied to the outside walls.
- A constant heat flux density (q) is imposed on the bottom base of the pins, with values given by experimental data.
- Symmetry planes are considered with no heat exchange.
- Conjugate heat transfer conditions were considered at the solid-fluid interface:

$$\begin{cases} q_w = -\lambda_f \dfrac{\partial T_f}{\partial n} = -\lambda_s \dfrac{\partial T_s}{\partial n} \\ T_f = T_s \end{cases} \tag{6}$$

where n is the vector normal to the boundary.

NB: The adiabatic assumption may introduce inaccuracies in cases where heat transfer through the walls is significant, such as in high-temperature applications or when the heat sink is in contact with other heat sources. However, for the specific application of cooling Synchrotron SOLEIL components, the adiabatic assumption is valid because the heat transfer through the walls is negligible compared to the heat dissipation through the mini-channels and pin-fins [33].

Figure 3 shows the geometry of heat sink model. It is displayed how the heat flux density is applied on the bottom wall.

4.3 Numerical procedures

The numerical calculations were carried out using the ANSYS FLUENT code with a finite volume method approach. The SIMPLE algorithm was used for pressure-velocity coupling. The second-order "Upwind" scheme was chosen for the discretization of the governing equations. Several turbulence models based on RANS were tested. The Realizable k-ε turbulence model was selected for this study due to its superior performance in capturing strong shear and separation flows, which are critical for accurately predicting the heat transfer and hydrodynamic behavior in mini-channel heat sinks with pin-fins. Compared to alternative turbulence models such as the Standard *k*-ε, *k*-ω, and SST k-ω models, the Realizable k-ε model provides better accuracy in flows with strong streamline curvature, vortices, and rotation, which are common in pin-fin arrays. The Standard k-ε model tends to overpredict turbulent kinetic energy in regions of strong shear, while the k-ω and SST k-ω models, although more accurate in near-wall regions, are computationally more expensive and less robust for the complex geometries considered in this study [30, 34, 35].

The modeled transport equations for k and ε in the Realizable k-ε model are:

$$\frac{\partial}{\partial t}(\rho k) + \frac{\partial}{\partial x_i}(\rho k u_i) = \frac{\partial}{\partial x_i}\left[\left(\mu + \frac{\mu_t}{\sigma_k}\right)\frac{\partial k}{\partial x_i}\right] + G_k + G_b - \rho\varepsilon - Y_M + S_k \tag{7}$$

$$\frac{\partial}{\partial t}(\rho \varepsilon) + \frac{\partial}{\partial x_i}(\rho \varepsilon u_i) = \frac{\partial}{\partial x_i}\left[\left(\mu + \frac{\mu_t}{\sigma_\varepsilon}\right)\frac{\partial \varepsilon}{\partial x_i}\right] + \rho C_1 S_\varepsilon - \rho C_2 \frac{\varepsilon^2}{k + \sqrt{\nu\varepsilon}} + C_{1\varepsilon}\frac{\varepsilon}{k} C_{3\varepsilon} G_b + S_\varepsilon \tag{8}$$

where: $G_k = \tau_{ij}\frac{\partial u_i}{\partial x_j} = \mu_t S^2$, $G_b = \beta g_i \frac{\mu_t}{\Pr_t}\frac{\partial T}{\partial x_i}$, $\beta = -\frac{1}{\rho}\left(\frac{\partial \rho}{\partial T}\right)_P$.

$$\tau_{ij} = \mu_t\left(2S_{ij} - \tfrac{2}{3}\tfrac{\partial u_k}{\partial x_k}\delta_{ij}\right) - \tfrac{2}{3}\rho k\delta_{ij} \ , \ \ S_{ij} = \tfrac{1}{2}\left(\tfrac{\partial u_i}{\partial x_j} + \tfrac{\partial u_j}{\partial x_i}\right) \ , \mu_t = \rho C_\mu \tfrac{k^2}{\varepsilon} C_1 = \ \max\left[0.43, \tfrac{\eta}{\eta+5}\right]$$

$where : \eta = S\frac{k}{\varepsilon}\ and\ S = \sqrt{S_{ij}S_{ij}}$ and C_μ is not constant as *RNG k-ε* model it is given by: $C_\mu = 1/A\left(_0 + A_S\frac{k}{\varepsilon}U^*\right)$.

A_0, A_S, C_2, $C_{1\varepsilon}$, $C_{3\varepsilon}$, σ_k and σ_ε are constants given in Ref. [36].

Here, G_k represents the generation of turbulent kinetic energy due to mean velocity gradients, G_b accounts for buoyancy effects, Y_M represents the contribution of fluctuating dilatation in compressible turbulence, and S_k and S_ϵ are user-defined source terms. The model constants C_1, C_2, σ_k, and σ_ϵ are calibrated to ensure accurate predictions of turbulent flows.

This numerical approach, combined with the selected turbulence model, ensures a robust and accurate simulation of the thermal and hydrodynamic behavior of the heat sink under investigation.

The choice of this model is not arbitrary, but it is extracted from several tests in bi-dimensional numerical simulations; the results will be the subject of another scientific paper. Conjugate conduction–convection heat transfers at the fluid-solid interface have also been considered. The physical model configuration is designed in accordance with existing experimental prototypes; and in order to reduce the computation time, the symmetry of the problem has been taken into account in the streamwise direction.

4.4 Grid independence study

An unstructured mesh has been deployed to present the internal complexity of the geometry, formed by a channel filled with a matrix of pins. The mesh has been adapted to assure satisfactory resolution coverage: close to the "adhesion" walls where the velocity gradients and temperature are expected to be large and to detect the flow velocity variation due to the decrease of the fluid passage cross-section between pins. Between 1 and 2 million nodes were used to generate the entire mesh of the computational domain.

A grid independence study was carried out to assure no influence of mesh on the simulation results. The grids were adopted based on the difference between the heat transfer coefficient and the fluid temperature and the effects between the two, which remained at less than 2%. The 2% threshold for grid independence was determined based on prior studies and engineering standards that consider deviations below 2% as acceptable for numerical simulations. This threshold ensures that the results are not significantly affected by the mesh resolution while maintaining computational efficiency [37].

The average fluid temperatures in the channel for a typical case are shown in **Table 1**. It can be seen that the average quantities deviations are about 1,631,060 nodes compared to the grids with 1,205,954 nodes are 0.3%, 0.1%, and 1.7%, respectively. The difference between these two grids is less than 2%. Furthermore, the size of the neighboring elements at the walls has been gently refined, introducing the improved near-wall treatment to ensure a better accurate resolution of any gradient in this region (**Figure 4**). It was confirmed that the mesh refinement is limited to a threshold to obtain independent solutions. Excessive grid refinement can lead to amplified truncation errors, resulting in physically meaningless solutions. This phenomenon is well-documented in the literature, where over-refinement can cause numerical instability and divergence of the solution [37]. We confirmed that the mesh refinement is limited to a threshold to get independent solutions. If we continue to refine, the truncation error will be amplified, and the solutions will no longer make sense physically.

Position	Boundary condition
z = 0	uniform heat density at the base
z = Δ and Δ + H	interface fluid-solid: conjugate heat transfer
z = 2Δ + H	adiabatic on top plan (x,y)
y = 0 and y = W	adiabatic on lateral plan (x,z)
x = 0 and x = L	adiabatic on plan (y,z)
x = 0	inlet velocity
x = L	outflow

Table 1.
Boundary condiction.

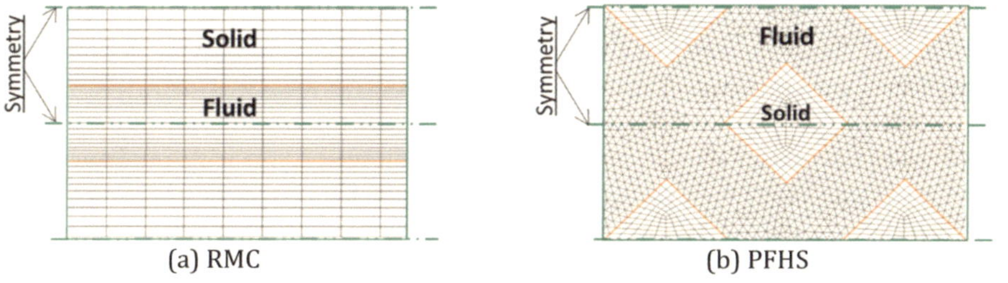

Figure 4.
Generated Mesh: (a) rectangular mini-channel and (b) diamond pin-fin.

Parameters	Grid D 233640	Error %	Grid E 326920	Error %	Grid F 434320
h_{ave} (W/m^{2}K)	3670	6.1%	3421	1.7%	3364
T_{ave} (K)	347.2	0.5%	345.6	0.3%	3465
T_{out} (K)	358.2	0.2%	357.6	0.1%	357.1

Table 2.
Grid independence calculations.

This rigorous grid independence study ensures the robustness and accuracy of the numerical simulations, providing confidence in the validity of the results (**Table 2**).

4.5 Parameters and data processing

The inlet Reynolds number was calculated for each simulation based on the prescribed mass flow rate. Special attention was given to estimating the Reynolds number in the fluid passage zones between the pin-fins, particularly focusing on the determination of the maximum velocity. The maximum velocity (U_{max}) is calculated using the following formula:

$$U_{\max} = \frac{\dot{m}}{\rho.A_{\min}} \tag{9}$$

The hydraulic diameter (D_h) is a critical parameter in determining the flow characteristics in mini-channels. In this study, the hydraulic diameter is calculated based on the rectangular cross-section of the fluid passage between the pins, with a height (H) and a width (e). This approach is commonly used for rectangular channels and is well-suited for the geometry of the pin-fin heat sink. For comparison, other methods of estimating the hydraulic diameter include:

1. *Circular cross-section*: For circular channels, the hydraulic diameter is simply the diameter of the channel. This method is straightforward but not applicable to non-circular geometries like the pin-fin heat sink.

2. *Annular cross-section*: For annular channels, the hydraulic diameter is calculated as the difference between the outer and inner diameters. This method is useful for concentric geometries but not relevant for the pin-fin configuration.

3. *Complex geometries*: For more complex geometries, the hydraulic diameter can be estimated using the ratio of the cross-sectional area to the wetted perimeter. This method is versatile but requires detailed geometric information.

The rectangular cross-section method used in this study provides a balance between accuracy and simplicity, making it suitable for the pin-fin heat sink geometry [33].

A specific methodology was employed to calculate U_{max}, which relies on the accurate estimation of the hydraulic diameter (D_h). It is worth noting that the coefficient factor in our approach differs from that of M. Liu [14] by approximately $\sqrt{2}$. Our method precisely follows the flow path between the pin-fins, considering a control surface of dimensions $e \times H$ (where e is the width of the fluid passage and H is the height), perpendicular to the flow direction, as illustrated in **Figure 2**. The maximum velocity is determined at the narrowest cross-section of the fluid passage within the pin-fin array.

The hydraulic diameter (D_h) is defined based on the rectangular cross-section of the fluid passage between the fins, with height (H) and width (e), and is expressed as:

$$\mathrm{Re} = \frac{U_{\max} D_h}{\nu};\ D_h = \frac{2eH}{e+H}.$$

A novel approach was developed in this study to compute the heat transfer coefficient, which is outlined as follows:

5. Local heat transfer coefficient

The local heat transfer coefficient at each face is defined as:

$$h(x) = \frac{q_w(x)}{T_w(x) - T_f(x)} \tag{10}$$

where

- $q_w(x)$ is the local heat flux density at the interface surface,
- $T_w(x)$ is the local temperature at the pin wall interface,
- $T_f(x)$ is the local temperature of the fluid at the channel axis opposite to the pin face.

6. Face-average heat transfer coefficient

The face-average heat transfer coefficient is calculated as:

$$h_{ave}(i) = \frac{q_{w,ave}(i)}{T_{w,ave}(i) - T_{f,ave}(i)} \tag{11}$$

where

- $q_{w,avg}(i)$ is the average heat flux density for a pin face,
- $T_{w,avg}(i)$ is the average temperature at the pin wall interface,
- $T_{f,avg}(i)$ is the average fluid temperature at the channel axis opposite to the pin face.

7. Global average heat transfer coefficient

The global average heat transfer coefficient for the entire channel is computed as:

$$\overline{h} = \frac{1}{N}\sum_{1}^{N} h_{ave}(i) \tag{12}$$

where

- $\overline{h}$ is the average heat transfer coefficient across all N faces,
- N is the number of pins in the streamwise direction.

Each pin has two faces (front and back), and the heat transfer coefficients for these faces are calculated separately. The front-side (h_{front}) and back-side (h_{back}) coefficients are determined by averaging the heat transfer coefficients of the top, bottom, and two lateral faces of each channel, as shown in **Figures 5** and **6**. The calculations are performed using the following expressions:

$$\begin{aligned} \overline{h}_{ave}(i)\big|_{front} &= \frac{1}{4}\left(h_{top,i} + h_{bottom,i} + 2h_{lateral,i}\right)\Big|_{front} \\ \overline{h}_{ave}(i)\big|_{back} &= \frac{1}{4}\left(h_{top,i} + h_{bottom,i} + 2h_{lateral,i}\right)\Big|_{back} \end{aligned} \tag{13}$$

This approach ensures a comprehensive and accurate evaluation of the heat transfer coefficient, accounting for the complex flow and thermal interactions within the pin-fin array. The methodology provides a robust framework for analyzing the thermal performance of the heat sink under varying flow conditions.

Both the local and average Nusselt numbers are defined using the hydraulic diameter (D_h) and the corresponding heat transfer coefficients (local and average). The Nusselt number is calculated as follows:

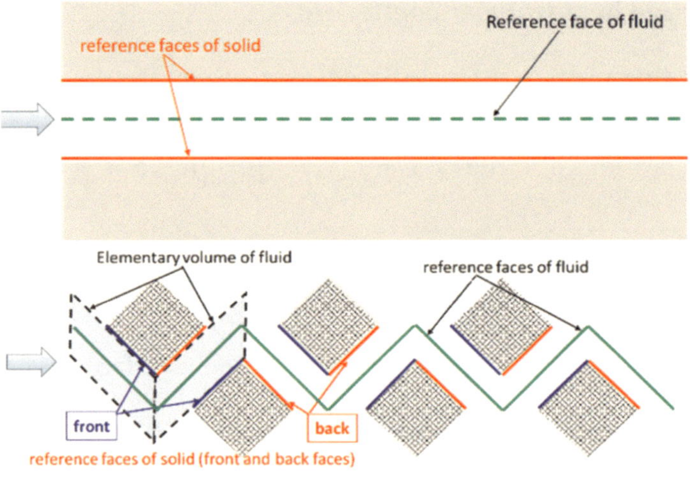

Figure 5.
Reference faces for fluid and solid zones.

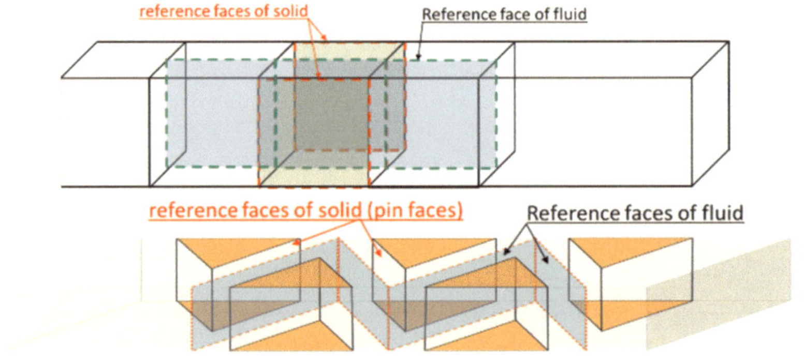

Figure 6.
Two channel configurations: (a) regular RMC (b) corrugated PFHS.

$$Nu_x = \frac{h_x D_h}{\lambda_x} Nu_{ave} = \frac{h_{ave} D_h}{\lambda_f} \tag{14}$$

where

- Nu_x is the local Nusselt number,
- Nu_{avg} is the average Nusselt number,
- h_x is the local heat transfer coefficient,
- h_{avg} is the average heat transfer coefficient,
- D_h is the hydraulic diameter,
- λ_f is the thermal conductivity of the coolant.

These equations establish the relationship between the heat transfer coefficient and the Nusselt number, providing a dimensionless measure of convective heat transfer efficiency in the system. The Nusselt number is a critical parameter for evaluating the thermal performance of the heat sink, as it quantifies the enhancement of heat transfer relative to conduction alone.

8. Results and discussion

Numerical simulations were conducted for two distinct configurations: a rectangular mini-channel (RMC) heat sink and a diamond-shaped pin-fin heat sink (PFHS). The results are presented for various mass flow rates and geometric dimensions, providing insights into the thermal and hydrodynamic performance of both configurations.

8.1 Validation of the numerical procedure

To ensure the accuracy and reliability of the numerical simulations, the results for pressure drop and heat transfer coefficient were compared with experimental data from the literature. Specifically, **Figure 7** compares the numerical pressure drop results for the diamond-shaped pin-fin heat sink (PFHS) with experimental data obtained at the Synchrotron SOLEIL by Rebay et al. [25]. Similarly, **Figure 8** presents a comparison of the numerical pressure drop results with experimental data from Liu et al. [14]. Both comparisons are made across a range of Reynolds numbers and fluid passage widths (e). In this context, "excellent agreement" is defined as a deviation of less than 5% between the numerical and experimental results. This threshold ensures that the numerical model accurately captures the physical behavior of the system and provides reliable predictions for further parametric studies.

Furthermore, the average heat transfer coefficient calculated using the numerical procedure was compared with experimental measurements from Rebay et al. [25]. As shown in **Figure 9**, a good agreement is reached for the PFHS case with $e = 0.35\ mm$.

However, the current validation is limited to specific geometries (diamond-shaped pin-fins and rectangular mini-channels) and flow conditions (mass flow rates ranging from 0.05 to 3 kg/min). To further strengthen the validity of the numerical model,

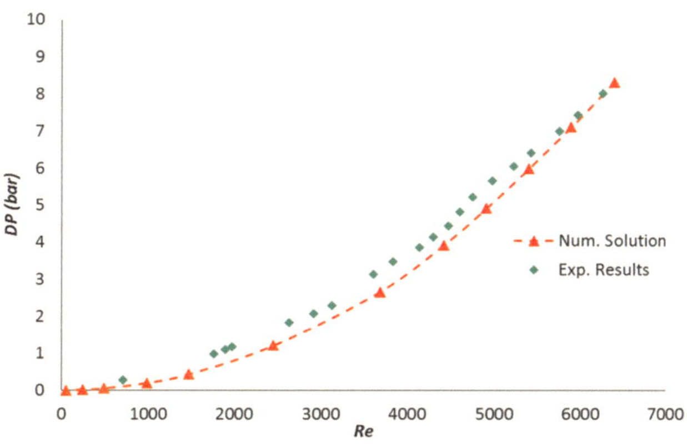

Figure 7.
Pressure drop versus Re (e = 0.35 mm) for PFHS compared to Exp. [25].

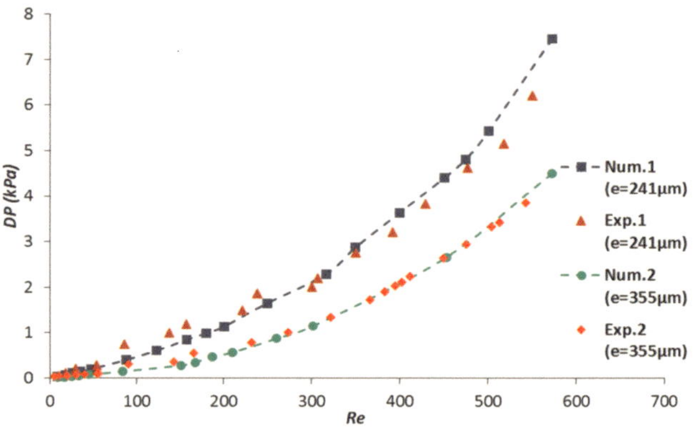

Figure 8.
Numerical pressure drop versus Re *for PFHS compared to Exp. [14].*

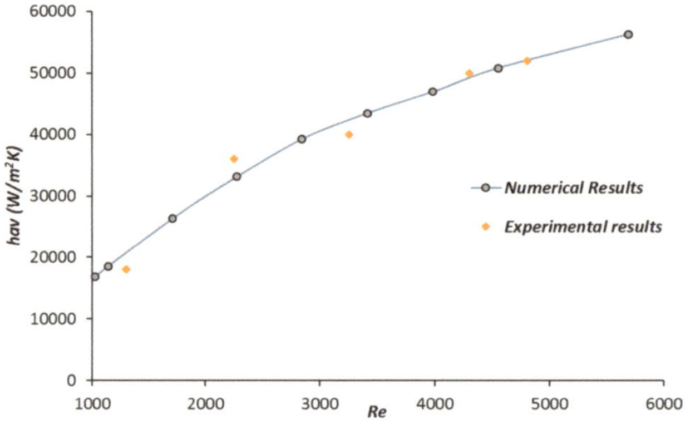

Figure 9.
Numerical averaged heat transfer coefficient compared to Exp. [25].

future studies could extend the validation to a wider range of geometries, such as different pin-fin shapes (e.g., circular and elliptical) and configurations (e.g., inline and staggered with varying pitch), as well as a broader range of flow conditions, including higher mass flow rates and different coolant fluids. This would provide a more comprehensive understanding of the model's applicability and robustness across diverse scenarios.

The present assessments highlight the robustness of the numerical methodology and its ability to accurately predict the thermal and hydrodynamic behavior of the heat sinks under study. The close agreement between the numerical and experimental results provides confidence in the reliability of the simulations and their applicability to real-world cooling systems.

From this comparative analysis, it can be concluded that the numerical simulation methodology employed in this study has been successfully validated. The results demonstrate good agreement with experimental data, confirming the accuracy and reliability of the computational approach. Consequently, this validation provides a solid foundation for conducting further parametric studies to explore additional aspects of the thermal and hydrodynamic performance of the heat sink under various operating conditions.

8.2 Results

The variation of the local heat transfer coefficient for the diamond-shaped pin-fin heat sink (PFHS) is illustrated in **Figure 10**. The profile exhibits a zigzag pattern, which arises due to the periodic alternation between the front and back sides of the pin-fins, as well as the discontinuities between adjacent pins. The peaks in the profile correspond to the front sides of the pins, where heat transfer is more pronounced, while the troughs represent the back sides, where heat transfer is relatively lower. This pattern is observed along a line of pin-fins in the streamwise direction. This zigzag pattern in the heat transfer coefficient profile is consistent with findings in the literature, where similar behavior has been observed in pin-fin arrays due to the periodic flow acceleration and deceleration around the pins [23].

The term "classic behavior" refers to the typical heat transfer coefficient profile observed in conventional heat sinks, characterized by periodic variations due to the pin-fin arrangement. These variations are caused by the alternating regions of flow acceleration (front sides of the pins) and flow separation (back sides of the pins), which lead to local maxima and minima in the heat transfer coefficient profile, respectively. It can also be seen that the classic appearance of the local heat transfer coefficient is found, in both cases of RMC and PFHS. This behavior aligns with established heat transfer principles and confirms the effectiveness of the pin-fin

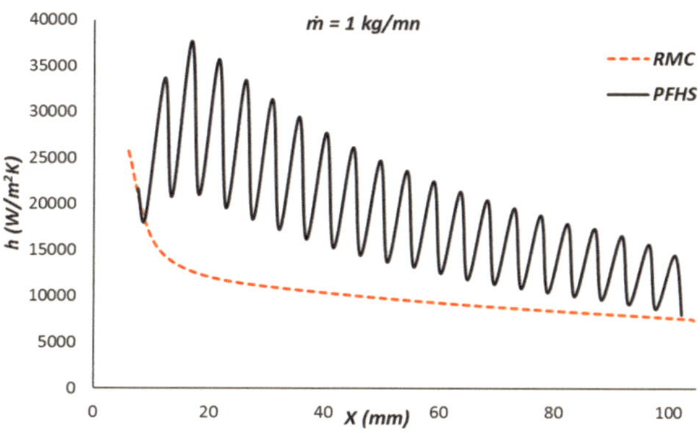

(a) e = 1.6mm, $\dot{m}$=1 kg/mn, Re_c =2300, Dh_c=1.6mm

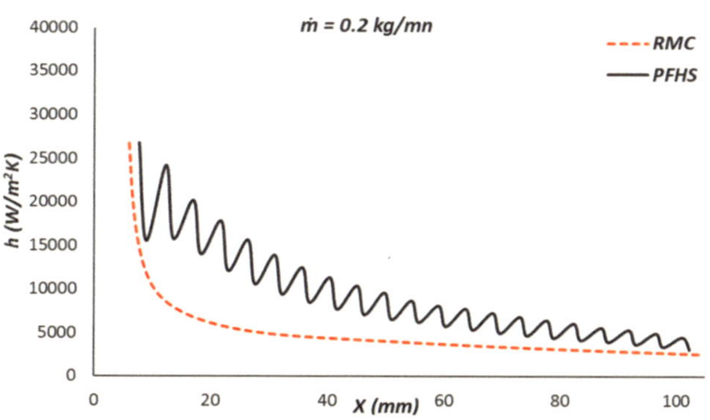

(b) e=1.6mm, $\dot{m}$ =0.2kg/mn , Re_c =132, Dh_c=1.6mm

Figure 10.
Local heat transfer coefficient along the PFHS and the RMC configuration (q = 100 kW/m2) (a) e = 1.6 mm, e = 1 kg/mn, Re_c = 2300, Dh_c = 1.6 mm; (b) e = 1.6 mm, e = 0.2 kg/mn, Re_c = 132, Dh_c = 1.6 mm.

design in enhancing thermal performance. The pin-fin heat sink (PFHS) demonstrates a significant improvement in heat transfer performance compared to the rectangular mini-channel (RMC). Specifically, the PFHS shows a 20% enhancement in the average heat transfer coefficient compared to the RMC under the same flow conditions. This improvement is attributed to the increased turbulence and better flow mixing induced by the staggered arrangement of the diamond-shaped pin-fins, which enhances heat dissipation and reduces thermal resistance [27].

Furthermore, the variable reference temperature (VRT) approach will be more illustrative than other methods, such as the constant bulk temperature approach, because it better captures local temperature variations along the flow path. Traditional methods that use a constant bulk temperature often fail to account for the significant temperature gradients that occur in complex geometries like pin-fin heat sinks, leading to inaccuracies in the heat transfer coefficient estimation. In contrast, the VRT approach dynamically adjusts the reference temperature based on local fluid temperatures, providing a more accurate representation of the temperature difference between the wall and the fluid. This results in a more precise calculation of the heat transfer coefficient, particularly in regions with strong temperature gradients or flow separation [38]. The effects of the reference temperature and its impact on heat transfer performance will be further discussed in the next section.

A fit function was developed to describe the variation of the averaged heat transfer coefficient (h_{avg}) with the Reynolds number (Re). The fit function is expressed as:

$$h_{ave} = a \cdot \mathrm{Re}^b + c \tag{15}$$

where

- a, b, and c are *fitting* coefficients determined from the numerical results,
- *Re* is the Reynolds number.

The corresponding error margins for the fit function are within ±5%, indicating a high level of accuracy in representing the numerical data. This fit function provides a useful tool for predicting the heat transfer performance of the pin-fin heat sink under different flow conditions [33].

The static temperature contours, depicted in **Figure 11**, reveal that the cooler regions of the fluid are consistently located in the central area of the channel. As the fluid progresses from the inlet to the outlet, this central region gradually heats up, causing the reference temperature to vary along the streamwise direction. This variation results in an increasing temperature difference between the fluid and the wall, reaching its maximum values along the channel. By selecting the middle faces as the reference area, lower fluid temperatures are maintained, ensuring a significant temperature difference ($\Delta T_m = T_w\text{-}T_f$). This difference is crucial for sustaining heat flux exchange at the wall until thermal equilibrium ($T_f = T_w$) is achieved.

Additionally, the middle region serves as an ideal zone for the effective mixing of hot and cold fluids. This area is characterized by prominent wake regions and accelerated flow, which enhance fluid mixing and promote efficient heat transfer. The flow dynamics in this region, as illustrated in **Figure 11**, highlight the importance of the middle faces in maintaining thermal stability and optimizing heat transfer performance.

DOI: http://dx.doi.org/10.5772/intechopen.1009791

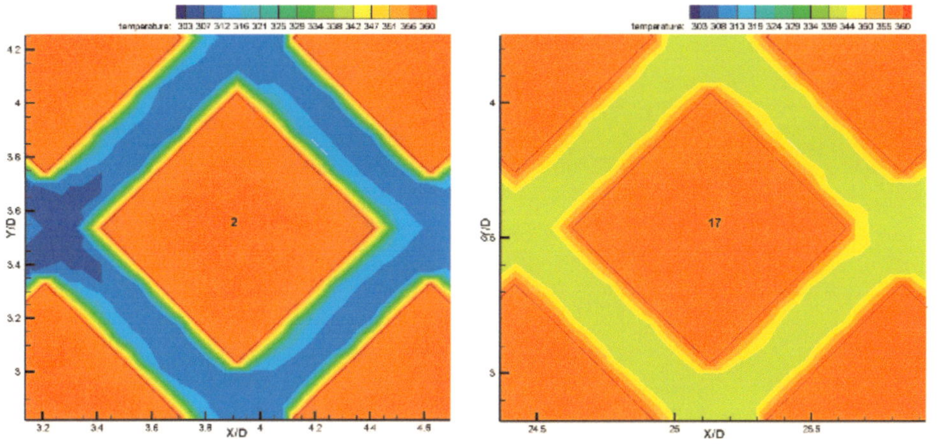

(a) 2nd pin (b) 17th pin

Figure 11.
Static temperature contours for PFHS (e = 0.7 mm).

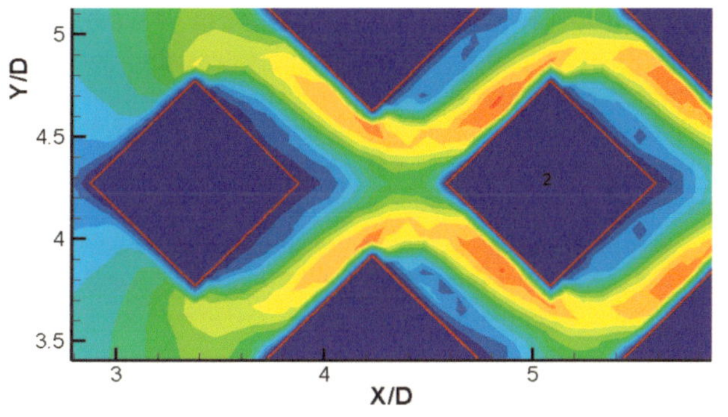

Figure 12.
Velocity magnitude contours (m/s) *with* (e = 1.2 mm *and* e = 1 kg/min).

Figure 12 displays contours of velocity magnitude for a low mass flow rate ($\dot{m}$ = 1 *kg/min*) at a fluid passage width of *e = 1.2 mm*. The flow exhibits periodic acceleration near the front edges of the pin-fins, particularly in the red regions, where the fluid velocity increases due to the narrowing of the flow passage. This acceleration is accompanied by the formation of significant recirculation zones, as clearly visible in the figure. The wake regions between the diamond-shaped pin-fins are prominent, contributing to enhanced mixing of hot and cold fluid streams. While this flow topology promotes heat transfer by improving thermal mixing, it also results in increased pressure drops, a trade-off inherent to such complex geometries.

Figure 13 illustrates the flow behavior in the staggered pin-fin arrangement, which generates higher flow resistance compared to an inline configuration. The recirculation zones are particularly pronounced behind the pin-fins, where the flow separates and elongates downstream. These recirculation regions, which begin after the wake, limit the mixing of the fluid shear layer in the back regions of the pin-fins. As a result, the heat transfer rate in these areas is reduced. The staggered arrangement, while effective in enhancing heat transfer through increased turbulence, also

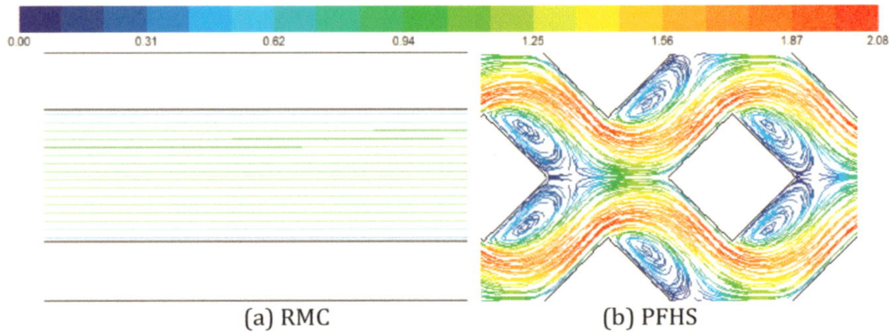

Figure 13.
Stream lines colored by velocity magnitude (m/s) *with* $\dot{m}$ = *1 kg/min.*

introduces challenges in maintaining uniform flow distribution and minimizing pressure losses.

In summary, the staggered diamond-shaped pin-fin configuration promotes heat transfer by inducing turbulence and recirculation, but it also leads to higher pressure drops and localized reductions in heat transfer efficiency in the wake regions. These findings highlight the importance of optimizing pin-fin geometry and arrangement to balance thermal performance and hydrodynamic efficiency.

In **Figure 13**, the staggered arrangement of pin-fins results in higher flow resistance compared to the inline arrangement. Notably, significant recirculation zones form behind the pin-fins, particularly in the back regions. These recirculation zones originate downstream of the wake region and extend further along the flow path. This flow behavior reduces the mixing of the fluid shear layer in the back regions, leading to a decrease in the local heat transfer rate. The staggered configuration, while enhancing flow resistance, creates complex flow patterns that impact the thermal performance of the heat sink.

8.3 Effects of fluid reference temperature

In this section, a refined approach is introduced to define the reference parameters used in calculating the heat transfer coefficient. These parameters are associated with reference faces for both the fluid and the pin walls, as illustrated in **Figure 5**. This method accounts for the variation in fluid temperature along the channel, leading to a more accurate evaluation of the heat transfer coefficient.

Figures 14 and **15** depict the temperature distribution under identical conditions for the rectangular mini-channel heat sink (RMC) and the diamond pin-fin heat sink (PFHS), respectively. The temperature profiles are shown on both vertical (xy) and horizontal (xz) planes at heights corresponding to 1%, 50%, and 99% of the channel height (H). It is evident that the temperature increases along the streamwise direction due to heat transfer from the base wall, which is subjected to a constant heat flux of $q = 100\ kW/m^2$. Additionally, the pin-fins play a crucial role in enhancing heat transfer by promoting the mixing of hot and cold fluid streams.

8.4 Fluid temperature and heat transfer analysis

The results also reveal that the lowest fluid temperatures are consistently found along the channel axis (fluid passage). This observation supports the use of the

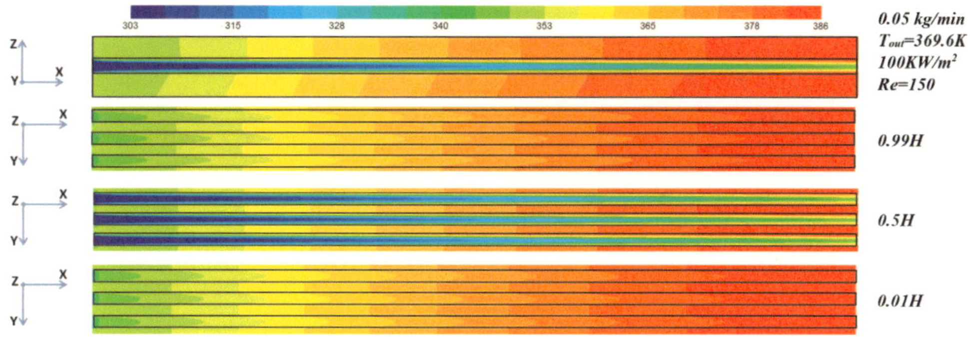

Figure 14.
*Temperature contour at longitudinal plan for RMC (*e = 1.6 mm*).*

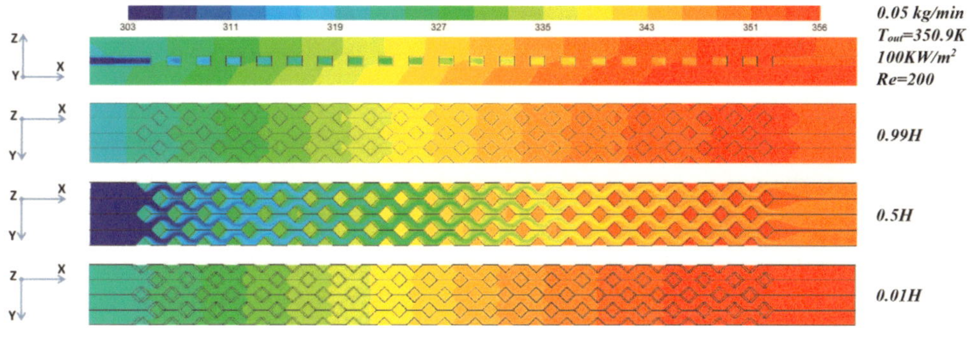

Figure 15.
Temperature contour at longitudinal plan for PFHS (e = 1.6 mm).

temperature difference ($\Delta T = T_w - T_f$) as a key parameter, where T_w is the wall temperature and T_f is the fluid temperature. The reference temperature ($T_{ref} = T_f$) is taken at the channel axis and varies from the inlet to the outlet, denoted as $T_{ref}(x)$. This approach is particularly suitable for evaluating the average heat transfer coefficient, as it captures the dynamic variation in fluid temperature along the channel. By adopting this method, the calculation of the heat transfer coefficient becomes more representative of the actual thermal behavior within the heat sink. This approach ensures a more accurate assessment of the heat transfer performance, particularly in complex geometries like the pin-fin heat sink.

The fluid temperature distribution in the xz plane (**Figure 14**) shows a noticeable increase from 303 *K* at the inlet to 369 *K* at the outlet for a mass flow rate of 0.05 *kg/min*, while the wall temperature reaches 356 *K*. In the *xy* plane at $y = 0.99H$, the temperature difference between the inlet and outlet, as well as between the wall and fluid at the outlet, is significant. Maintaining a consistent temperature difference along the heat sink is crucial for optimal performance. The highest temperature difference is observed at higher mass flow rates in the mini-channel configuration.

For the pin-fin heat sink (**Figure 15**), the fluid temperature in the xz plane rises from *303 K* to *350 K* at a mass flow rate of 0.05 *kg/min*, with the wall temperature reaching *356 K*. In the xy plane at $y = 0.99H$ and $y = 0.5H$, the temperature difference between the inlet and outlet, as well as between the wall and fluid at the outlet, remains significant. Improved cooling performance is observed in the *xy* plane at $y = 0.5H$, where the temperature distribution is more uniform.

8.5 Estimation of thermal performance

The accurate estimation of the heat sink's thermal performance relies on the proper calculation of the average heat transfer coefficient, which is strongly dependent on the reference fluid temperature (T_{ref}). Several approximations for T_{ref} were evaluated:

1. Inlet Temperature: $T_{ref} = T_{in}$
2. Average Temperature of the Channel Volume: $T_{ref} = T_{bulk}$
3. Arithmetic Mean of Inlet and Outlet Temperatures: $T_{ref} = (T_{out} + T_{in})/2$
4. Local Average Temperature Along the Channel: $T_{ref} = T_{ave}(x)$

Figure 16 compares the fluid temperature profiles along the channel for these reference temperature approaches. It is evident that both $T_{out\,-\,in}$ (arithmetic mean of inlet and outlet temperatures) and T_{bulk} (average channel volume temperature) start with values higher than the wall temperature. However, the inlet fluid temperature should ideally be lower than the wall temperature to ensure effective cooling. As the wall temperature increases along the channel, it intersects the $T_{out\,-\,\mathrm{in}}$ and T_{bulk} profiles, resulting in $\Delta T = 0$ at these points. This leads to indeterminate values of the heat transfer coefficient, rendering these approximations unsuitable.

When the reference temperature is set to T_{in}, the wall-fluid temperature difference (ΔT) increases continuously from the inlet to the outlet. However, this behavior contradicts the expected trend, as ΔT should decrease along the channel due to heat transfer from the wall to the fluid. Therefore, the approximations $T_{out\,-\,in}$, T_{in}, and T_{bulk} are not appropriate for accurate heat transfer coefficient calculations.

In contrast, when the reference temperature is defined as the local average fluid temperature along the channel axis ($T_{ref} = T_{ave}(x)$), the temperature difference (ΔT_m) is highest at the inlet and gradually decreases as the wall and fluid temperatures approach each other. A low ΔT_m value indicates efficient cooling, validating the effectiveness of this approach. This method provides a more accurate representation of the heat transfer process, ensuring reliable thermal performance evaluation.

The local heat transfer coefficient is presented in **Figure 17** for various reference temperature approaches. It is clear that the use of $T_{out\,-\,in}$ (arithmetic mean of inlet

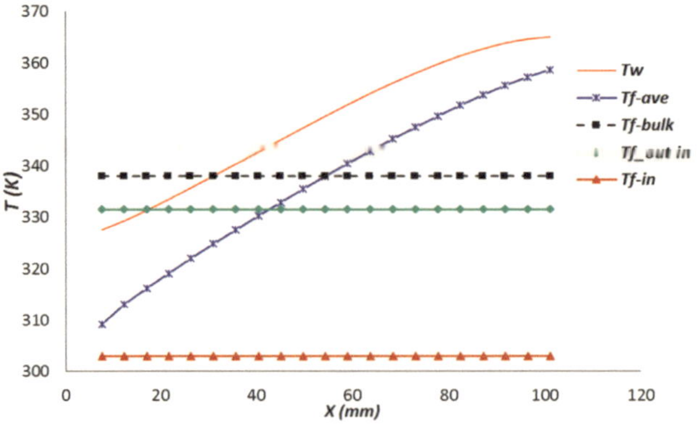

Figure 16.
Comparison of reference temperature methods for PFHS (e = 1.6 mm and $\dot{m}$ = 0.04 kg/min).

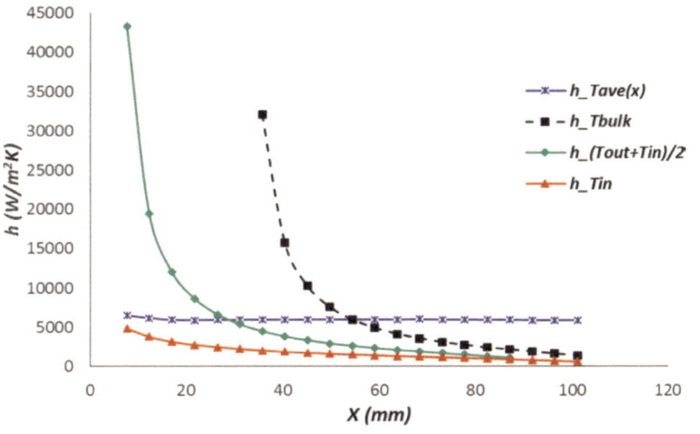

Figure 17.
Comparison of heat transfer coefficients (e = 1.6 mm, q = 100 kW/m² and $\dot{m}$ = 0.04 kg/min).

and outlet temperatures) and T_{bulk} (bulk fluid temperature) leads to unsatisfactory results, with discontinuities observed at $x = 17$ and $x = 31$, respectively. These discontinuities correspond to points where the temperature difference (ΔT) between the wall and the fluid becomes zero, rendering the heat transfer coefficient undefined. Additionally, the local heat transfer coefficient calculated using $T_{out\,-\,in}$, T_{bulk}, and T_{in} (inlet temperature) approaches tends to exhibit very low values near the heat sink outlet, which is inconsistent with expected thermal behavior.

In contrast, the heat transfer coefficient profile calculated using the variable reference temperature approach ($T_{ave}(x)$) aligns well with established literature. This profile starts with a high value at the inlet and gradually decreases to an asymptotic value as the flow reaches a thermally developed regime. This behavior is physically consistent, as it reflects the transition from an initial high heat transfer rate at the inlet to a stabilized thermal condition downstream. The $T_{ave}(x)$ approach, which considers the local average fluid temperature along the channel, provides a more accurate and reliable representation of the heat transfer coefficient, avoiding the discontinuities and unrealistic trends observed with other methods.

Figures 18–21 illustrate the distribution of fluid and wall heat flux, as well as local temperatures, for the pin-fin heat sink with fluid passage dimensions of e = 1.6 mm and

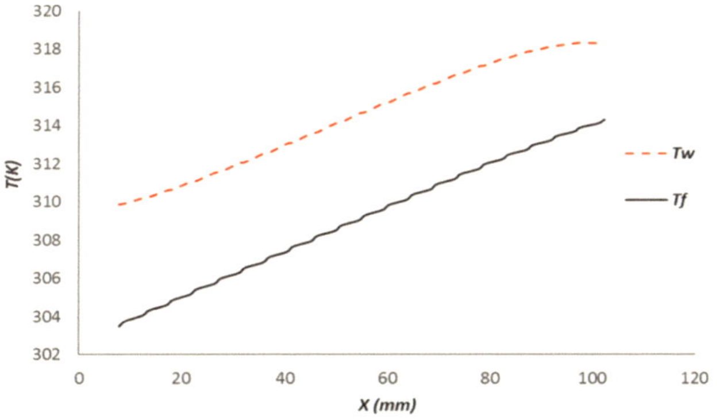

Figure 18.
Fluid and wall temperature (e = 1.6 mm and $\dot{m}$ = 0.2 kg/min).

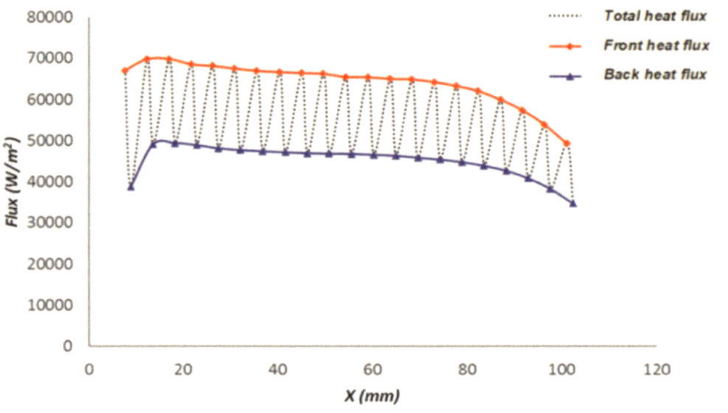

Figure 19.
Total, front and back heat flux (e = 0.35 mm and $\dot{m}$ = 0.2 kg/min).

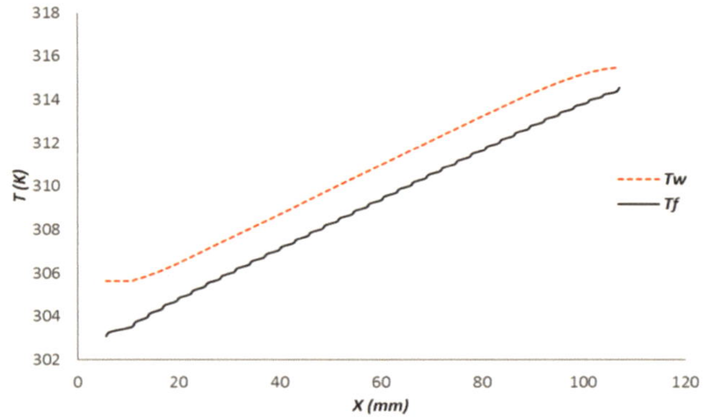

Figure 20.
Fluid and wall temperature (e = 0.35 mm and $\dot{m}$ = 0.2 kg/min).

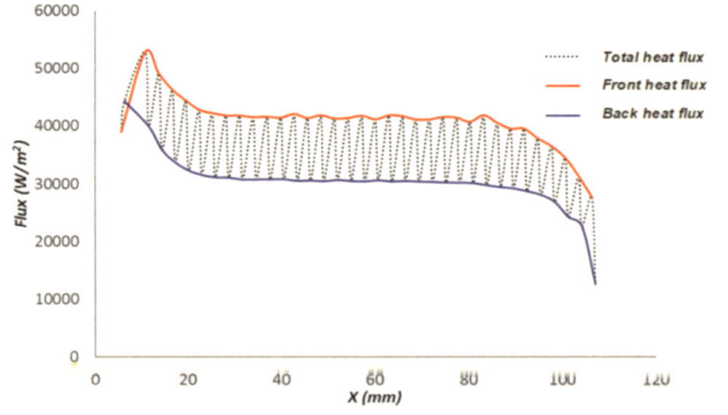

Figure 21.
Total, front and back heat flux (e = 0.35 mm and $\dot{m}$ = 0.2 kg/min).

e = 0.35 *mm*. The local temperature is determined by averaging the temperatures along the front and back fluid passage axes, as depicted in the reference faces shown in **Figure 5**.

The results reveal an intermittent pattern in the total heat flux along the heat sink, particularly between the front and back faces of the pin-fins. Higher heat flux values are observed on the front faces of the pins, where the fluid flow is more direct and intense. In contrast, lower heat flux values are found on the back faces, which are less exposed to the fluid flow due to the presence of recirculation zones, as shown in **Figure 12(b)**. These recirculation zones, characterized by slower and less turbulent flow, are less effective for convective heat transfer, leading to reduced thermal performance in these regions.

This analysis highlights the importance of pin-fin geometry and fluid flow dynamics in optimizing heat transfer efficiency, particularly in complex configurations where recirculation and flow separation can significantly impact thermal performance.

Figure 22 illustrates the local heat transfer coefficient profiles for the bottom, top, and lateral faces of both the rectangular mini-channel (RMC) and the pin-fin heat sink (PFHS). These calculations were conducted using the variable reference temperature (VRT) approach, with a constant applied heat flux of $q = 100\ kW/m^2$ and a mass flow rate of $\dot{m} = 1\ kg/min$.

For the rectangular mini-channel, the heat transfer coefficient profiles for all faces (bottom, top, and lateral) exhibit similar behavior. The profiles start with high values at the inlet and gradually decrease, stabilizing to constant values toward the outlet, which corresponds to the establishment of a well-developed thermal regime.

In contrast, the pin-fin heat sink displays distinct behavior. The heat transfer coefficient begins with relatively low values at the inlet and increases along the length of the heat sink, eventually reaching a constant value near the outlet. This trend is attributed to the vortices generated by the pin-fins, which enhance heat transfer as the flow progresses. Additionally, the lateral surfaces of the pin-fin heat sink exhibit

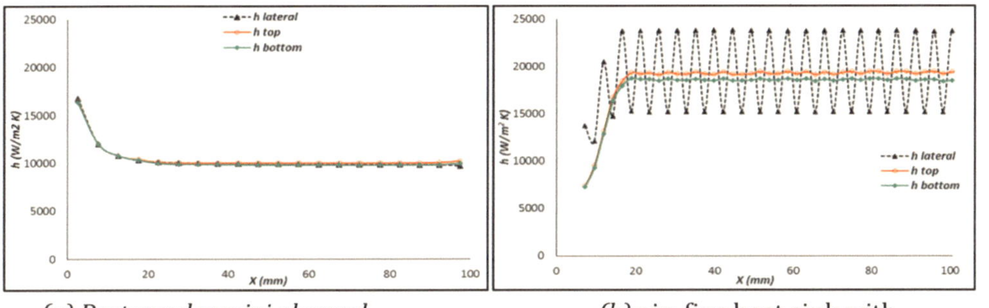

(a) Rectangular mini-channel *(b)* pin-fins heat sink with

Figure 22.
Local heat transfer coefficient for e = 1.6 mm and $\dot{m}$ = 1 kg/min.

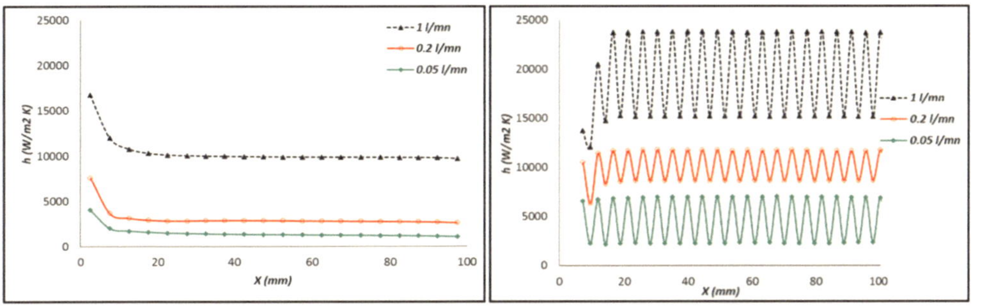

(a)RMC *(b)PFHS*

Figure 23.
Heat transfer coefficient for different flow rates (q = 100 kW/m2 and e = 1.6 mm).

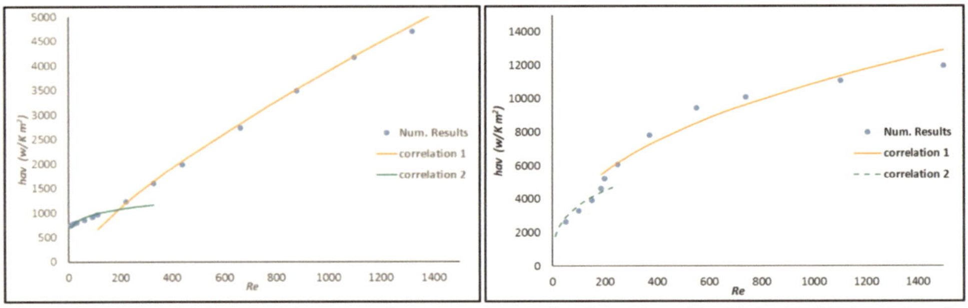

h_{ave} based on $T_{ave}(x)$ for RMC ($e=1.6mm$)
correlation 1: $h_{ave}= 17\ Re^{0,780}$
correlation 2: $h_{ave}= 504\ Re^{0,143}$

h_{ave} based on $T_{ave}(x)$ for PFHS ($e=1.6mm$)
correlation 1: $h_{ave}= 634\ Re^{0,412}$
correlation 2: $h_{ave}= 870\ Re^{0,310}$

Figure 24.
Local heat transfer coefficient for different Reynolds numbers.

strong oscillations in the heat transfer coefficient compared to the top and bottom surfaces, reflecting the complex flow dynamics induced by the pin-fin geometry.

These observations highlight the differences in thermal performance between the rectangular mini-channel and the pin-fin heat sink, emphasizing the influence of geometric complexity on heat transfer behavior.

Figure 23 illustrates the local heat transfer coefficient profiles for varying mass flow rates, calculated using the proposed fluid reference temperature approach. The heat transfer coefficient (h_w) was determined for the lateral faces of both the rectangular mini-channel (RMC) and the diamond-shaped pin-fin heat sink (PFHS). As anticipated, the results demonstrate that the heat transfer coefficient increases with higher mass flow rates, corresponding to an increase in the Reynolds number. This trend aligns with established principles of convective heat transfer, where enhanced fluid flow leads to improved thermal performance. The findings highlight the effectiveness of the proposed approach in capturing the relationship between flow dynamics and heat transfer in complex geometries.

8.6 Averaged heat transfer coefficient fit function

A fit function was developed for the two configurations studied in this work to express the averaged heat transfer coefficient as a function of the Reynolds number, using the reference temperature approach $T_{ave}(x)$. As shown in **Figure 8**, the heat transfer coefficients obtained from this approach demonstrate good agreement with the experimental measurements reported by Rebay et al. [25]. The fit functions are presented in a simplified form, making them practical for engineering applications. These proposed functions are illustrated in **Figure 24**, where it is evident that they accurately represent the numerical results. This formulation provides a reliable tool for predicting heat transfer performance in similar systems.

9. Conclusion

This study presents a numerical investigation of flow and heat transfer in rectangular mini-channels (RMC) and diamond-shaped pin-fin heat sinks (PFHS). A detailed analysis was conducted to evaluate the influence of fluid reference temperature variation on the calculation of the heat transfer coefficient in such complex geometries. The methodology for determining the reference temperature was critically examined, and various evaluation methods were compared. The numerical procedure was validated against experimental measurements, demonstrating excellent agreement. The key findings of this study are summarized as follows:

1. Heat transfer coefficient profile: A distinct zigzag pattern in the heat transfer coefficient profile was observed, resulting from the alternating front and back faces of the pin-fins. Additionally, the classic behavior of the local heat transfer coefficient was consistently observed in both the RMC and PFHS configurations.

2. Fluid temperature distribution: The coldest regions of the fluid were consistently located at the center of the channel. This ensures that a significant temperature difference between the wall and the fluid is maintained along the channel, facilitating continuous heat transfer from the inlet to the outlet.

3. Thermal performance comparison: The thermal performance of the pin-fin heat sink (PFHS) was found to be superior to that of the rectangular mini-channel (RMC), even at lower mass flow rates, while maintaining acceptable pressure drop levels.

4. Variable reference temperature (VRT) Approach: The proposed variable fluid reference temperature (VRT) approach effectively replicates the concept of the well-known T_∞, representing the fluid temperature far from the wall. This method provides a more accurate representation of the heat transfer process.

5. Advantages of the VRT approach: The VRT approach ensures a consistent and physically realistic profile of the heat transfer coefficient, offering a clearer understanding of heat transfer enhancement. In contrast, other methods for calculating the reference temperature tend to overestimate the heat transfer coefficient, particularly at the inlet, and underestimate it at the outlet, which contradicts the expected physical behavior of pin-fin heat sinks.

6. Design implications: The diamond-shaped pin-fin configuration plays a critical role in enhancing the cooling capacity of the heat sink. The thermal performance of this complex geometry is highly sensitive to the accurate calculation of the heat transfer coefficient. The variation in fluid temperature is identified as a key factor in determining the heat transfer coefficient (h), which in turn influences the overall performance of the heat sink.

To the best of the authors' knowledge, this work represents the first rigorous effort to estimate the heat transfer coefficient in such a complex geometry. Future studies will focus on developing correlations for the heat transfer coefficient to further refine the design and optimization of pin-fin heat sinks.

Nomenclature

A	cross-sectional area of fluid
C	passage
Cp	specific heat (J/kg K)

Dh	hydraulic diameter (m)
e	dimension of fluid passage
f	friction coefficient
H	pin/channel height
h	convective heat transfer coefficient
L	dissipater length
$\dot{m}$	mass flow rate (*kg/min)*
N	number of pins
Nu	Nusselt number
P	absolute pressure (Pa)
q	heat flux density (W/m^2)
Re	Reynolds number
T	temperature (K)
t	temperature fluctuation (K)
ΔT_m	mean temperature difference wall-fluid
U	fluid velocity (m/s)
u	velocity fluctuation (m/s)
W	heat sink width (m)
x,y,z	specials coordinates (m)
ΔP	pressure drop across heat sink (Pa)

Greek symbols

ρ	density (kg/m^3)
μ	dynamic viscosity (kg/m.s)
ν	kinematic viscosity (m^2/s)
λ	thermal conductivity (W/m K)

Subscripts

ave.	average
f	fluid
i, j	subscript of
in	inlet
out	outlet
bulk	averaged quantity over domain
ref	reference
w	wall
c	channel
p	pin

Abbreviation

RMC	rectangular mini-channel
PFHS	pin-fin heat sink
VRT	variable reference temperature

Author details

Nabil Bessanane
LESEI Laboratory, University of Batna2, Faculty of Technology, Batna, Algeria

*Address all correspondence to: n.bessanane@univ-batna2.dz

References

[1] Zukauskas AA. Heat transfer from tubes in cross flow. In: Advances in Heat Transfer. Vol. 8. New York: Academic Press; 1972. pp. 93-160

[2] Moores KA, Joshi YK. Effect of tip clearance on the thermal and hydrodynamic performance of a shrouded pin fin array. Journal of Heat Transfer. 2003;**125**(16):999-1006

[3] Shkarah AJ, Sulaiman MYB, Ayoba MRBH, Togun H. A 3D numerical study of heat transfer in a single-phase micro-channel heat sink using graphene, aluminum and silicon as substrates. International Communications in Heat and Mass Transfer. 2013;**48**:108-115

[4] Obot NT. Toward a better understanding of friction and heat/mass transfer in microchannels—A literature review. Microscale Thermophysical Engineering. 2002;**6**:155-173

[5] Kandlikar SG, Grande WJ. Evolution of microchannel flow passages-thermohydraulic performance and fabrication technology. Heat Transfer Engineering. 2003;**24**:3-17

[6] Kandlikar SG, Grande WJ. Evaluation of single-phase flow in microchannels for high flux chip cooling e thermo-hydraulic performance enhancement and fabrication technology. Heat Transfer Engineering. 2004;**25**(18):5-16

[7] Bahrami M, Yovanovich MM, Culham JR. Pressure drop of fully developed, laminar flow in rough microtubes. Journal of Fluids Engineering. 2006;**128**:632-637

[8] Kosar A, Peles Y. Thermal-hydraulic performance of MEMS-based pin fin heat sink. Journal of Heat Transfer. 2006;**128**(12):121-131

[9] Weilin Q, Mudawar I. Experimental and numerical study of pressure drop and heat transfer in a single-phase micro-channel heat sink. International Journal of Heat and Mass Transfer. 2001; **45**:2549-2565

[10] Kosar A, Mishra C, Peles Y. Laminar flow across a bank of low aspect ratio micro pin fins. Journal of Fluids Engineering. 2005;**127**(13):419-430

[11] Qian M, Mei DQ, Yi ZD, Feng YB, Chen ZC. Fluid flow and heat transfer performance in a micro-reactor with non-uniform micro-pin-fin arrays for hydrogen production at low Reynolds number. International Journal of Hydrogen Energy. 2017;**42**(1):553-561. DOI: 10.1016/j.ijhydene.2016.10.150

[12] Morini GL. Single-phase convective heat transfer in microchannels: A review of experimental results. International Journal of Thermal Sciences. 2004;**43**: 631-651

[13] Jiang PX, Xu RN. Heat transfer and pressure drop characteristics of mini-fin structures. International Journal of Heat and Fluid Flow. 2007;**28**:1167-1177. DOI: 10.1016/j.ijheatfluidflow.2006.11.008

[14] Liu MH, Liu D, Xu S, Chen YL. Experimental study on liquid flow and heat transfer in micro square pin fin heat sink. International Journal of Heat and Mass Transfer. 2011;**54**(25–26): 5602-5611. DOI: 10.1016/j.ijheatmasstransfer.2011.07.013

[15] Yang J, Wang L, Li HZ. 2-dimensional CFD simulation and correlation development for optimizationof fin heatsinks in electronic cooling. Journal of Thermal Science.

2001;**10**:363-371. DOI: 10.1007/s11630-001-0045-2

[16] Mebarki G, Rahal S. Passive control of twophase two phase flow thermal instabilities in a vertical tube evaporator. Journal of Thermal Science and Engineering Applications. 2016;**8**(4):041008. DOI: 10.1115/1.4034092

[17] Tullius JF, Tullius TK, Bayazitoglu Y. Optimization of short micro pin fins in minichannels. International Journal of Heat and Mass Transfer. 2012;**55**(15–16): 3921-3932. DOI: 10.1016/j.ijheatmasstransfer.2012.03.022

[18] Yang DW, Wang Y, Ding GF, Jin ZY, Zhao JH, Wang GL. Numerical and experimental analysis of cooling performance of single-phase array microchannel heat sinks with different pin-fin configurations. Applied Thermal Engineering. 2017;**112**:1547-1556. DOI: 10.1016/j.applthermaleng.2016.08.211

[19] Eren M, Caliskan S. Effect of grooved pin-fins in a rectangular channel on heat transfer augmentation and friction factor using Taguchi method. International Journal of Heat and Mass Transfer. 2016;**102**:1108-1122. DOI: 10.1016/j.ijheat masstransfer.2016.07.005

[20] Babar H, Hongwei W, Ali HM, Zhang W. Hydrothermal performance of inline and staggered arrangements of airfoil shaped pin-fin heat sinks: A comparative study. Thermal Science and Engineering Progress. 2023;**37**:101-616

[21] Hasan MI. Investigation of flow and heat transfer characteristics in micro pin fin heat sink with nanofluid. Applied Thermal Engineering. 2014;**63**(2): 598-607. DOI: 10.1016/j.applthermaleng.2013.11.059

[22] Xu JL, Gan YH, Zhang DC, Li XH. Microscale heat transfer enhancement using thermal boundary layer redeveloping concept. International Journal of Heat and Mass Transfer. 2005; **48**(9):1662-1674. DOI: 10.1016/j.ijheatmasstransfer.2004.12.008

[23] Chyu MK, Hsing YC, Natarajan V. Convective heat transfer of cubic fin arrays in a narrow channel. Journal of Heat Transfer. 1998;**120**(2):362-367. DOI: 10.1115/1.2841414

[24] Chyu MK, Hsing YC, Shih TP, Natarajan. Heat transfer contributions of pins and endwall in pin-fin arrays: Effects of thermal boundary condition modeling. Journal of Turbomachinery. 1999;**121**(2):257-263

[25] Rebay M, Kakac S, Cotta RM, editors. Microscale and Nanoscale Heat Transfer: Analysis, Design, and Application. 1st ed. Boca Raton: CRC Press; 2016. DOI: 10.1201/b19261

[26] Chyu MK, Hsing YC, Natarajan V. Experimental and numerical analysis of heat transfer in pin-fin arrays with different spacing configurations. International Journal of Heat and Mass Transfer. 1999;**42**(20):3781-3791. DOI: 10.1016/S0017-9310(99)00034-6

[27] Garimella SV, Wirtz RA. Heat transfer and pressure drop in pin-fin arrays with various spacing ratios. Journal of Heat Transfer. 2001;**123**(3): 567-575. DOI: 10.1115/1.1370502

[28] Bar-Cohen A, Iyengar M. Optimization of pin-fin heat sinks for enhanced thermal performance. IEEE Transactions on Components and Packaging Technologies. 2002;**25**(2): 309-317. DOI: 10.1109/TCAPT.2002.1010005

[29] Soodphakdee D, Behnia M, Copeland DW. A comparison of fin

geometries for heat sinks in laminar forced convection: Part 1 e round, elliptical and plate fins in staggered and inline configuration. International Journal of Electronics and Microcircuits. 2001;**24**(1):First Quarter

[30] Wilcox DC. Turbulence Modeling for CFD. 3rd ed. La Cañada, CA: DCW Industries; 2006

[31] Pope SB. Turbulent Flows. Cambridge: Cambridge University Press; 2000

[32] Versteeg HK, Malalasekera W. An Introduction to Computational Fluid Dynamics: The Finite Volume Method. 2nd ed. Harlow: Pearson Education; 2007

[33] Incropera FP, DeWitt DP, Bergman TL, Lavine AS. (1994), Fundamentals of Heat and Mass Transfer. 7th ed. Hoboken, NJ: John Wiley and Sons; 2013

[34] Menter FR. Two-equation eddy-viscosity turbulence models for engineering applications. AIAA Journal. 1994;**32**(8):1598-1605. DOI: 10.2514/3.12149

[35] Shih TH, Liou WW, Shabbir A, Yang Z, Zhu J. A new k-ε eddy viscosity model for high Reynolds number turbulent flows. Computers and Fluids. 1995;**24**(3):227-238. DOI: 10.1016/0045-7930(94)00032-T

[36] A. T. R. a. M. P. L. Davis. A Comparison of RANS-Based Turbulence Modeling for Flow over a Wall-Mounted Square Cylinder, N C Motorsports and Automotive Research Center, Department of Mechanical Engineering and Engineering Science. Charlotte, NC, USA: The University of North Carolina at Charlotte; 2012

[37] Ferziger JH, Perić M. Computational Methods for Fluid Dynamics. 3rd ed. Berlin: Springer; 2002

[38] Moffat RJ. Describing the uncertainties in experimental results. Experimental Thermal and Fluid Science. 1988;**1**(1):3-17. DOI: 10.1016/0894-1777(88)90043-X

Chapter 4

High-Temperature Air Effects in High-Speed Flows

Konstantin Volkov

Abstract

A numerical simulation is conducted to model the flow of chemically reacting air at equilibrium conditions using a high-temperature air effect. The results are then compared to those obtained from a perfect gas model. The mathematical formulation of the high-temperature air model is provided, and it is verified and validated using computational fluid dynamics benchmark problems such as flow around a forward step and flow around a sphere. The accuracy of numerical calculations of fluid quantities in high-speed flows is addressed. The flow patterns and shock wave structures computed with the inviscid fluid model, the model accounting for temperature-dependent viscosity, and the model accounting for turbulent effects are compared. Various options for simulation of high-temperature processes in air are assessed. The shape and location of the bow shock wave and the occurrence of numerical solution fluctuations behind the shock (wave front) are highlighted with an emphasis on the impact of different factors.

Keywords: aerodynamics, shock wave, high-temperature air, perfect gas, sphere, forward step

1. Introduction

In contrast to subsonic flows, supersonic flows exhibit sharp changes and disruptions in flow properties. Precisely simulating supersonic flows to replicate actual physical phenomena necessitates employing high-accuracy methods that minimize numerical viscosity to levels below physical viscosity or eliminate numerical dissipation, as mentioned in Ref. [1]. Implementing low-accuracy schemes results in smoothing the flow pattern due to numerical effects, causing it to deviate from the real flow structure, as stated in Ref. [2]. The chosen numerical method's accuracy in resolving complex gas-dynamic structures is of computational interest.

When a high-intensity shock wave moves, air behind its front experiences high temperatures and pressures. The perfect gas model is insufficient for providing accurate numerical solutions in such cases because the constants for molecular weight and heat capacities are dependent on pressure and temperature [3]. In this case, processes of dissociation and ionization occurring in high-temperature air play an important role. In practical applications, various models that include high-temperature effects in air are utilized, including analytical relationships and interpolation of tabular data. One particularly interesting and effective model for air is the one presented in Ref. [4]

(Kraiko model), which takes into account the reactions between 13 components. This model includes ionization and dissociation processes of air. The model error typically falls within ±1% interval and does not exceed ±2% in the temperature interval up to 2×10^4 K and pressure interval from 10^{-3} to 10^3 atm. The flow around hypersonic aircraft (waverider) is successfully simulated using the Kraiko model [5].

The Euler or Navier–Stokes equations, which at supersonic flow velocities are of hyperbolic type, cannot be numerically solved unless the condition of non-decreasing entropy is satisfied [6]. It turns out that the non-decreasing entropy condition cannot be satisfied by simply adhering to the conservation laws. To solve the entropy problem, artificial viscosity is introduced. The distribution of gas-dynamic discontinuities across multiple computational mesh cells is impacted by the appearance of artificial viscosity.

The Godunov method, which finds widespread use in the numerical solution of Euler equations, offers a monotonic numerical solution. The exact solution of the Riemann problem is a key element of the Godunov method. Scheme viscosity fills the artificial viscosity role in the Godunov method. Godunov's method provides end-to-end calculation of flows with gas-dynamic discontinuities with a minimum scheme viscosity and first-order accuracy. Approximation techniques are used to solve the Riemann problem due to the complexity of exact solution [7].

The Roe scheme uses the fluid quantities on faces of the control volume to find numerical fluxes. The flux through face of the control volume corresponds to linearized solution of the Riemann problem in the configuration with two shock waves and fulfillment of the relations on single discontinuities. The scheme has proven itself well in flow simulations with shock waves and contact discontinuities; however, it has known entropy problems (appearance of non-physical jumps) in calculating rarefaction zones when the characteristics change sign. To eliminate this shortcoming, various entropy correction procedures have been developed related to the addition of dissipation [6].

Rusanov scheme proposed in Ref. [8] is an effective finite-difference scheme for the simulation of two-dimensional flows with the introduction of explicit artificial viscosity. Rusanov's method is based on a two-wave scheme, taking into account discontinuities such as left and right shock waves without considering the contact discontinuity. Rusanov scheme is based on the introduction of terms with artificial diffusion of a general form, and forward differences in time and central differences in spatial variables are taken. Rusanov's scheme uses a local maximum for parameters in adjacent mesh cells. Thus, not only is artificial viscosity introduced into the scheme, but also artificial thermal conductivity and artificial mass diffusion.

Advection Upstream Splitting Method (AUSM) family of schemes represents a convective wave and an acoustic wave as two physically different processes [9]. The inviscid flux is divided into two parts: convective flux and pressure flux. AUSM scheme shows good accuracy in flow simulations with strong shock waves. ENO/WENO schemes, although they have good accuracy in the simulation of a wide range of supersonic gas flows, are characterized by rather high implementation complexity on unstructured grids and increased requirements for computing resources [10, 11].

The results of a few Computational Fluid Dynamics (CFD) benchmark problems using the perfect air and high-temperature air models are presented in this study. Several finite-difference schemes are used to discretize convective fluxes. The outcomes of numerical computations using the high-temperature air model and the perfect gas model are contrasted with information found in published works.

DOI: http://dx.doi.org/10.5772/intechopen.1010657

2. Governing equations

Conservative equations of mass, momentum, and energy in integral form are as follows:

$$\frac{\partial}{\partial t}\int_V U dV + \int_{\partial V} F dS = 0, \tag{1}$$

where t is the time, V is the control volume with the boundary ∂V, S is the surface area, n is the vector of outer normal to the boundary ∂V, U is the vector of conservative variables at point x at time t, and F is the tensor flux. The vector of conservative variables and tensor flux are

$$U = \begin{pmatrix} \rho \\ \rho v \\ \rho e \end{pmatrix}, F = \begin{pmatrix} \rho v \\ \rho v v + pI - \tau \\ (\rho e + p)v - v\tau + q \end{pmatrix}. \tag{2}$$

Total energy is the sum of internal energy (including vibrational, rotational, and translational energy as well as electronic excitation of air atomic and molecular components) and kinetic energy.

$$e = \varepsilon + \frac{1}{2}|v|^2. \tag{3}$$

The variables are density, ρ, pressure, p, velocity vector, v, with components v_x, v_y, and v_z in coordinate directions x, y, and z, viscous stress tensor τ, and unit tensor I. The viscous stress components are

$$\tau_{ij} = \mu\left(\frac{\partial v_i}{\partial x_j} + \frac{\partial v_j}{\partial x_i} - \frac{2}{3}\frac{\partial v_k}{\partial x_k}\delta_{ij}\right). \tag{4}$$

The Fourier law is applied to find the heat flux vector.

$$q - \lambda \nabla T. \tag{5}$$

Viscosity and thermal conductivity are two essential properties of air. Viscosity and thermal conductivity are solely functions of temperature and are essentially independent of pressure for a thermally perfect gas. The dependencies of these properties on flow quantities become more complex in high-temperature air when processes of dissociation and ionization are apparent. The pressure dependence becomes significant (because of a change in the molecular makeup). The Sutherland formulas are used to calculate dynamic viscosity.

$$\mu = \mu_0 \frac{C + T_0}{C + T}\left(\frac{T}{T_0}\right)^{3/2}, \tag{6}$$

where $\mu_0 = 1.81 \times 10^{-5}$ kg/(m s) at $T_0 = 273$ K, $C = 110.4$ K.

The equation of state is $p = (\rho R_0 T)/M(p, T)$, where M is the molecular weight of air, which typically varies with temperature and pressure, and R_0 is the universal gas constant. When enthalpy, $h = \varepsilon + p/\rho$, is introduced, total energy is

$$e = h + \frac{1}{2}|v|^2 - \frac{p}{\rho}. \tag{7}$$

The enthalpy is defined as

$$h = h_0 + \int_{T_0}^{T} c_p dT, \tag{8}$$

where h_0 represents the substance formation enthalpy at $T = T_0$.

For a perfect gas, the Mayer relationship holds that $c_p - c_v = R$ and $\gamma = c_p/c_v$ which is the formula for the ratio of specific heat capacities. In this case, $h = c_p T$ and $\varepsilon = c_v T$ yield the enthalpy and internal energy, respectively.

3. Finite volume method

On unstructured meshes, the Navier–Stokes equations are solved numerically with the finite volume method and the Godunov method. Polyhedra of any shape make up mesh cells. The mesh quantity is defined at the center of the control volume V_i. This value represents the fluid quantity averaged over the control volume.

$$U_i = \frac{1}{V_i} \int_{V_i} U dV. \tag{9}$$

Eq. (1) is presented in semi-discrete form. The integral over the boundary of ith control volume is the sum of the products of the numerical fluxes at the centers of j th face and area of the face, S_{ij}. Then

$$\frac{dU_i}{dt} + \frac{1}{V_i} \sum_j^{N_i} F_{ij} S_{ij} = 0, \tag{10}$$

where F_{ij} is flux from ith cell to jth cell, S_{ij} is area of jth face of ith control volume, and V_i is volume of ith control volume.

Using the explicit third-order Runge–Kutta scheme, the time derivative in Eq. (1) is discretized. There are several methods for calculating convective fluxes through control volume faces. Some conventional schemes for calculating flux, such as the Roe scheme, result in divergence of the computational process and loss of accuracy. Godunov and Rusanov are two discretization techniques for convective fluxes. Interpolation, which limits the gradient of solution from the cell center to the face of the control volume, achieves a second order of approximation in space and guarantees scheme monotonicity. An explicit scheme is used to approximate the viscous part of the flux.

Numerical fluxes are found as

$$F_{j+1/2} = \frac{1}{2}[F(U_L) + F(U_R)] - \frac{1}{2}|A|(U_R - U_L), \tag{11}$$

where the diagonal matrix $|A| = R|\Lambda|L$, $\Lambda = diag\{v_n - a, v_n, v_n + a\}$, has Jacobian eigenvalues as its elements.

The entire problem of modeling high-velocity flows can be broken down into discrete subtasks using a methodology developed for the effective heat capacity ratio, which allows us to account for high-temperature effects. The governing Eq. (1) can be computed numerically using methods that were designed for a perfect gas. The effective adiabatic index is introduced to extend these methods to flows of chemically reacting equilibrium air, allowing the perfect gas formula to be used to determine the speed of sound $a^2 = \gamma$. There is a relationship between the effective adiabatic exponent γ and the ratio of specific heat capacities ($\gamma = c_p/c_v$) is

$$\gamma = \gamma \frac{\rho}{p\rho_p}. \tag{12}$$

The speed of sound is

$$a^2 = \left(\frac{\partial p}{\partial \rho}\right)_s = \left(\rho_p + \rho_T \frac{dT}{dp}\right)^{-1}. \tag{13}$$

At thermodynamic equilibrium, the equality $Tds = dh - dp/\rho$ is valid. Since $dh = h_p dp + h_T dT$, then (if $ds = 0$)

$$\frac{dT}{dp} = \frac{1 - \rho h_p}{\rho h_T}. \tag{14}$$

Second-order center-difference discretization, or numerical differentiation, is used to calculate specific heat capacity at constant pressure.

$$c_p(p, T) = \left(\frac{\partial h}{\partial T}\right)_p = \frac{h(p, T + \delta T) - h(p, T - \delta T)}{2\delta T}, \tag{15}$$

where $\delta T = 0.01T$.

The Godunov method with exact solutions of the Riemann problem or Roe approximate solution, combined with Neumann artificial viscosity, eliminates decreased entropy in the solution of hyperbolic equations.

4. Transformation of variables

Four independent variables in the Euler equations are

$$U = \{\rho, \rho u_x, \rho u_y, \rho e\}, \tag{16}$$

where ρ is density, ρu_x and ρu_y are momentum projections on axes x and y, $\rho e = \rho\left[\varepsilon + \left(u_x^2 + u_y^2\right)/2\right]$ is total energy.

There are other sets of variables used in addition to the conservative set. This set is represented by physical variables

$$Q = \{\rho, u_x, u_y, p\}, \tag{17}$$

where ρ is density, u_x and u_y are velocity components in x and y directions, and p is pressure.

The molecular weight, M_Σ, the gas constant $R = R_0/M_\Sigma$, the heat capacity at constant pressure, c_p, and the heat capacity at constant volume, c_v, of a perfect gas are constant. Therefore, Mayer's relationship ($c_p - c_v = R$) is applicable. The ratio of specific heat capacities is $\gamma = c_p/c_v$, and the relations $h = c_pT$ and $\varepsilon = c_vT$ are applied to find the enthalpy and internal energy.

Conservative variables can be used to express physical variables and vice versa. The reverse transition necessitates the solution of governing equations, whereas the direct transformation operation is not difficult.

The expression for total energy and the equation of state make up the first set of equations for the transition:

- Step 1

$$\rho e = \rho\varepsilon(p,T) + \rho\frac{v^2}{2};$$
$$p = \rho\frac{R_0}{M(p,T)}T, \tag{18}$$

- Step 2

$$\varepsilon(p,T) = \frac{(\rho e)}{\rho} - \frac{v^2}{2};$$
$$pM(p,T) - \rho R_0 T = 0. \tag{19}$$

The unknowns are p, T, and $\varepsilon = h - p/\rho$. There is non-linearity in these equations. The Newton–Raphson method is applied to solve them, whereby the values of the variables at each iteration are determined from

$$p^{n+1} = p^n + \Delta p, T^{n+1} = T^n + \Delta T, \tag{20}$$

where Δp, ΔT are increments of unknown functions. Index n denotes the iteration number.

Only first-order terms remain after equations are linearized by expanding the functions ε and M in a Taylor series expansion

$$\varepsilon^{n+1}(p,T) = \varepsilon^n + \frac{\partial\varepsilon^n}{\partial p}\Delta p + \frac{\partial\varepsilon^n}{\partial T}\Delta T;$$
$$M^{n+1}(p,T) = M^n + \frac{\partial M^n}{\partial p}\Delta p + \frac{\partial M^n}{\partial T}\Delta T. \tag{21}$$

When the expressions for temperature and pressure are substituted into governing equations and terms of order two and higher are removed, the result is

$$\frac{\partial\varepsilon^n}{\partial p}\Delta p + \frac{\partial\varepsilon^n}{\partial T}\Delta T = \frac{(\rho e)}{\rho} - \frac{v^2}{2} - \varepsilon^n;$$
$$\left(M^n + p^n\frac{\partial M^n}{\partial p}\right)\Delta p + \left(p^n\frac{\partial M^n}{\partial T} - \rho R_0\right)\Delta T = \rho R_0 T^n - p^n M^n. \tag{22}$$

The increments of Δp and ΔT are found by solving linear equations. When the increments Δp and ΔT meet the required accuracy condition, the iterative process comes to an end. The first approximation is based on the solution at the previous time layer.

The derivatives of the functions ε and M are computed using the central difference derivative of second order of accuracy

$$\begin{aligned}
\frac{\partial \varepsilon^n}{\partial p} &= \frac{\varepsilon(p^n + \delta_p, T^n) - \varepsilon(p^n - \delta_p, T^n)}{2\delta_p}; \\
\frac{\partial \varepsilon^n}{\partial T} &= \frac{\varepsilon(p^n, T^n + \delta_T) - \varepsilon(p^n, T^n - \delta_T)}{2\delta_T}; \\
\frac{\partial M^n}{\partial p} &= \frac{M(p^n + \delta_p, T^n) - M(p^n - \delta_p, T^n)}{2\delta_p}; \\
\frac{\partial M^n}{\partial T} &= \frac{M(p^n, T^n + \delta_T) - M(p^n, T^n - \delta_T)}{2\delta_T}.
\end{aligned} \tag{23}$$

Here, $\delta_p = k_p p^n$ and $\delta_T = k_T T^n$. Coefficients k_p and k_T are selected from the interval 0.005–0.01.

5. Results and discussion

The example of solving several CFD benchmark problems with exact or reference numerical solutions demonstrates the capabilities of the developed numerical simulation tools. The outcomes of computations made using the ideal and high-temperature air models are contrasted with one another and with data from published sources.

5.1 Flow around the forward step

Validation of the developed algorithms involved calculating the planar channel flow with a rectangular step. The geometry is shown in **Figure 1**. The computational domain is a suddenly narrowed channel. The dimensions of the channel are three lengths and one height in the inlet section (all relative units). Beginning at the left edge and moving 0.6 distance to the right boundary, a 0.2-height step is made.

Gas enters the channel from the left at supersonic speed, creating a detached shock wave in front of the step that is sporadically reflected in the form of a triple configuration from the channel's upper wall. This test problem enables the analysis of the Prandtl–Meyer rarefaction flows and flows with shock waves. The channel length is $L = 3$ m, its height is $h = 1$ m, the step height is $h_s = 0.2$ m, and the step offset from the entrance is $x_s = 0.6$ m. This geometry is equivalent to the test problem presented

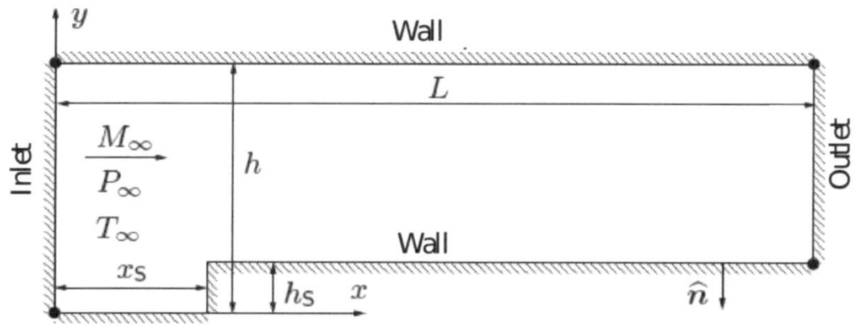

Figure 1.
Computational domain.

in Ref. [12]. However, it is suggested that the input gas-dynamic parameters be adjusted to match the parameters found in the atmosphere at a height of 30–40 kilometers. At this altitude during flight, propulsion systems may encounter such flows in flat channels.

The Kraiko model for air at high temperatures and perfect gas with a ratio of specific heat capacities $\gamma = 1.4$ are calculated. The static pressure, $p = 287$ Pa, the static temperature, and the Mach number ($M = 3.5$) are set on the inlet boundary. The free outflow conditions are applied to the outlet boundary of the channel. Since the flow within the outlet section is supersonic, the outlet boundary conditions have no effect on the flow structure within the computational domain. Boundaries reflecting mirror images are applied to the upper and lower edges.

At low flow pressures and intense shock waves, areas of non-physical solution [13] may appear in the numerical schemes. To suppress such effects, artificial viscosity can be introduced into the scheme.

Ten thousand nodes make up the mesh used for the computations. A convergence criterion is the minimum residual level. In the computations, the residual level ($R = 10^{-10}$) is reached after about 2200 time steps. Calculation options for the Roe and Rusanov schemes with and without viscous effects are investigated. The accuracy criteria for the numerical solution are the length of the upper Mach disk, the triple point (point of intersection of the upper Mach disk and bow shock), and the interaction between shock waves and upper walls of the step and channel [12].

When supersonic flow strikes a step and is reflected off the step and the channel walls, it creates a shock wave. Bell-shaped shock wave is produced after the step reflection. The Kelvin-Helmholtz instability arises at the contact discontinuity, where the Mach disk appears as a result of the interaction of this shock wave with the upper wall. The evaluation of calculation techniques is based on these structures' resolution quality. Non-physical features of the flow include the formation of a weak rarefaction wave when a fan of rarefaction waves originating from the step interacts with the step's upper wall and Kelvin-Helmholtz-type instability that propagates along the upper wall from the triple point. The small entropy oscillations brought on by numerical schemes at the triple point (computational boundary layer) are the main factor responsible for the onset and development of instability.

The solution is characterized by a significant non-monotonicity of the numerical solution and a distortion of the flow topology, according to the calculations carried out in Refs. [14, 15] using Ansys Fluent. When compared to the data presented in Ref. [12], there is a 24% overestimation of the Mach disk length, an 8% displacement of the triple point to the left in the axial direction, and 8% and 6% shifts in the points where oblique shock waves interact with step and channel upper walls. To avoid problems caused by the existence of a singular point, a circular arc with a small radius of 0.01 is used in place of the corner point.

The contours of velocity and temperature when calculated by the Roe scheme at zero viscosity ($\mu = 0$) are shown in **Figure 2**. The flow field is characterized by the presence of non-physical fluctuations, and the flow itself is unstable.

The modulus of velocity and temperature contours, which were calculated using the Roe scheme and non-zero gas viscosity, are displayed in **Figure 3** ($\mu = 1.8 \times 10^{-5}$ Pa s). These correspond to the viscosity of air at normal pressure and temperature. Although the calculation results show that there are slight fluctuations in the gas-dynamic parameters, the flow is more stable in this instance.

Figure 2.
Velocity contours (a) and temperature contours (b) computed with Roe scheme (μ=0).

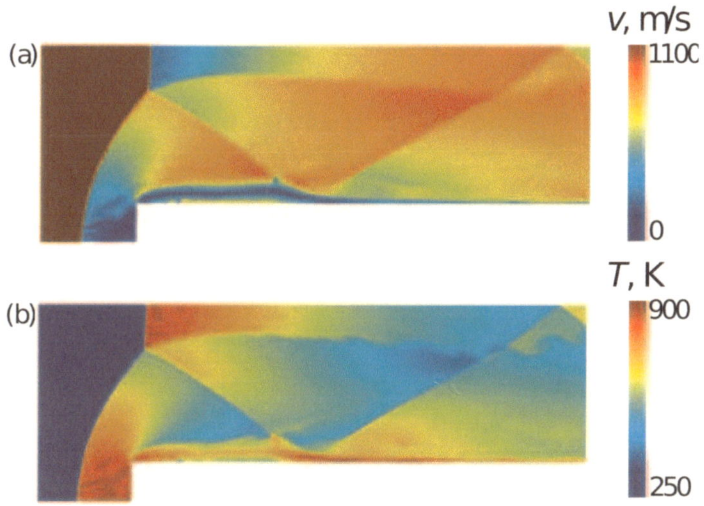

Figure 3.
Velocity contours (a) and temperature contours (b) computed with Roe scheme (μ=1.8×10^{-5}).

With a further increase in viscosity, for example, by a factor of 1000 up to $\mu = 1.8 \times 10^{-3}$ Pa s, the distributions of flow quantities become smoother, and fluctuations in flow quantities are observed only behind a direct shock wave (**Figure 4**).

Rusanov scheme is more dissipative in comparison with the Roe scheme, but it allows us to obtain a solution to the problem at low pressures without an additional increase in viscosity to non-physical values. In this case, there are no non-physical fluctuations of velocity and temperature, and the numerical solution is monotonic (**Figure 5**). The weak non-monotonicity of the numerical solution takes place only on

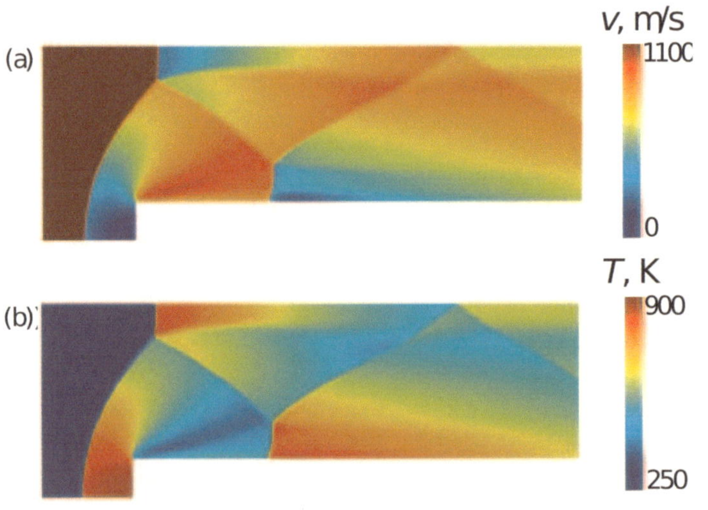

Figure 4.
Velocity contours (a) and temperature contours (b) computed with Roe scheme ($\mu = 1.8 \times 10^{-3}$).

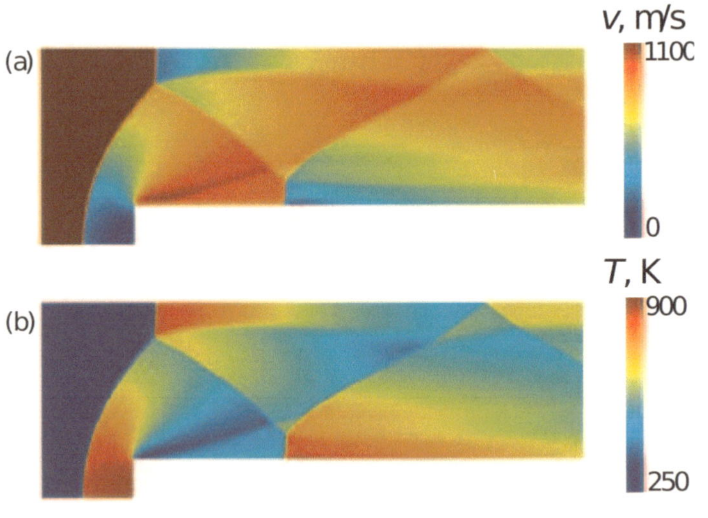

Figure 5.
Velocity contours (a) and temperature contours (b) computed with Rusanov scheme ($\mu = 0$).

a coarse mesh in the region between the step and the reflected shock wave. The average discrepancy between the numerical solution and the reference solution [12] does not exceed 3%. At the same time, there is no numerical noise observed in Refs. [16, 17] on the upper surface of the step.

The results obtained show that the Rusanov scheme is more suitable for calculating flows at low static gas pressure, although it is more dissipative. At moderate flow rates, there are no flow regions with high air temperatures. The use of perfect gas and the Kraiko thermodynamic model of high-temperature air [4] leads to almost identical results.

5.2 Flow around a sphere

Hypersonic flow around a sphere is investigated experimentally in Refs. [18-20], and the Kraiko model implementation on general-purpose GPUs is validated in Refs. [21, 22]. The discretization of convective fluxes using the Roe and AUSM schemes is compared in the calculations. The gradient in the cell centers is computed using the least squares method, and values are interpolated to the face from the cell centers to increase the order of approximation of the finite-difference scheme. To guarantee that the plane is monotonous, gradient limiters are employed.

An analysis is conducted on a hypersonic airflow around a $D = 12.7$ mm sphere with equilibrium chemical reactions. The sphere-to-bow shock wave distance is calculated using selected flow quantities that agree with the experiment results [19]. The free flow pressure is $p_\infty = 666.61$ Pa, and the temperature is $T_\infty = 293$ K. The density is $\rho_\infty = 7.9 \times 10^{-3}$ kg/m^3 given these values. The Mach number varies in interval $M_\infty = 7.10$–17.77 as flight speed is $V_\infty = 2438.4$–6705.6 m/s.

Figure 6 displays the sketch of the computational domain. At the stagnation point and the top point, the computational domain's outer boundary is separated from the sphere by 4 and 8.65 mm, respectively. An axisymmetric formulation of the problem is found and solved. The boundary conditions of supersonic inflow are set on the inlet boundary, while supersonic outflow conditions are applied to the outlet boundary. The wall has no-slip boundary conditions, but to facilitate setting the output boundary condition at the end of the sphere, a small section of the wall is designated as a non-viscous wall with slip. The surface of the sphere is thought to be thermally insulated. There are $400 \times 400 = 160000$ hexagonal cells in the computational grid.

For a methodological study of the developed calculation method, flow around a sphere is simulated at Mach number $M = 12$. The calculation results obtained with the AUSM scheme in an inviscid formulation are shown in **Figure** 7. The presented results show that behind the shock, there are anomalies and flow instability. There are strong instabilities behind the shock in the pressure field. Significant irregularities are also observed in the pressure distribution over the sphere. These effects are studied in

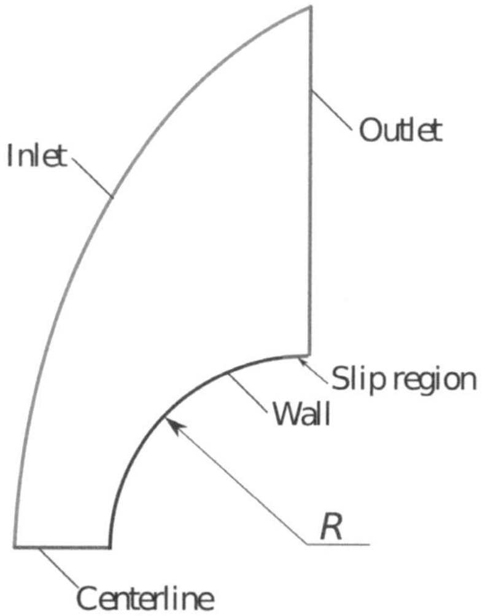

Figure 6.
Computational domain.

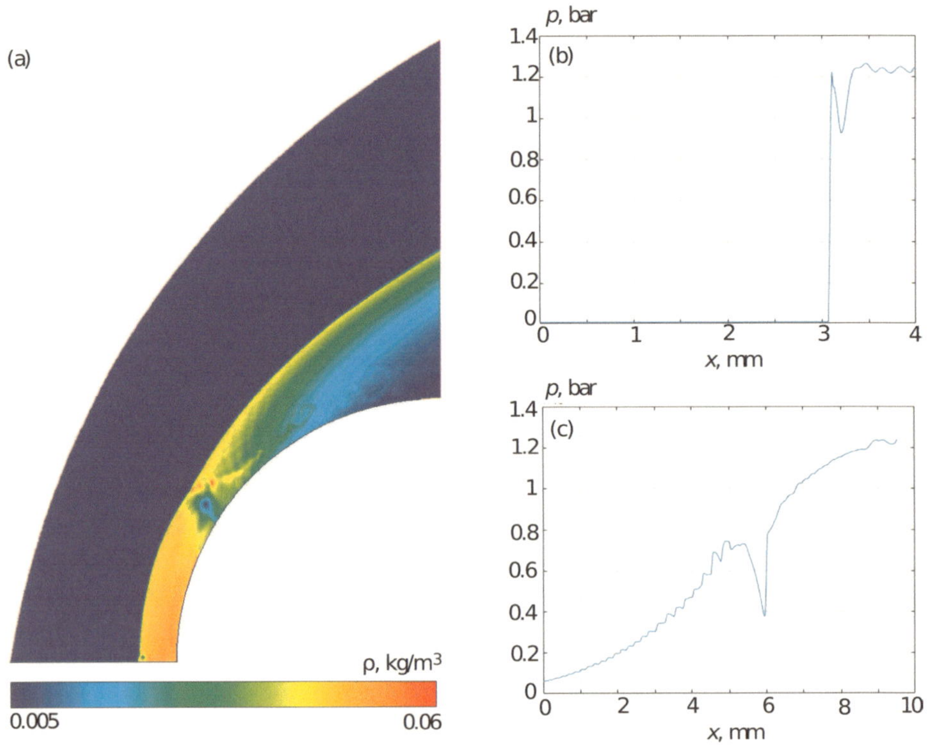

Figure 7.
Density contours (a), centerline pressure distribution (b), and surface pressure distribution (c) computed with the AUSM scheme and inviscid model.

detail, for example, in Ref. [13] and are related to the instability of the numerical scheme when calculating problems with high speed and low pressure.

The inclusion of laminar viscosity in the finite-difference scheme adds stability. The problem is complicated by the fact that the computational domain is chosen in such a way that both subsonic and supersonic flows are realized in the boundary layer when viscous flow is on the outlet boundary near the surface. This makes it difficult to correctly supply boundary conditions on the inlet boundary. In this regard, a boundary condition with slippage is set on a part or the entire surface of the sphere. At the same time, at the outlet, the flow near the sphere accelerates to supersonic velocities, which makes it possible to use the boundary condition of supersonic outflow at the outlet boundary. To reduce the area of influence of the slip boundary conditions, it is set only on a small part of the sphere surface, which is located near the outlet boundary. On the rest of the sphere surface, the boundary condition for tangential velocity and the thermal insulation condition for temperature are specified.

The density distribution obtained using the Roe scheme is presented in **Figure 8** and shows that there is a non-physical form of the bow shock. The Roe scheme is inoperable at high speeds and low pressures. A stable solution can only be obtained with a very small Courant number of the order of 0.01, which leads to a significant expenditure of processor time.

Numerical simulations are carried out in the computational domain shown in **Figure 6**. The shape and size of the domain are identical in CFD simulations.

Figure 8.
Density contours computed with the Roe scheme and inviscid model.

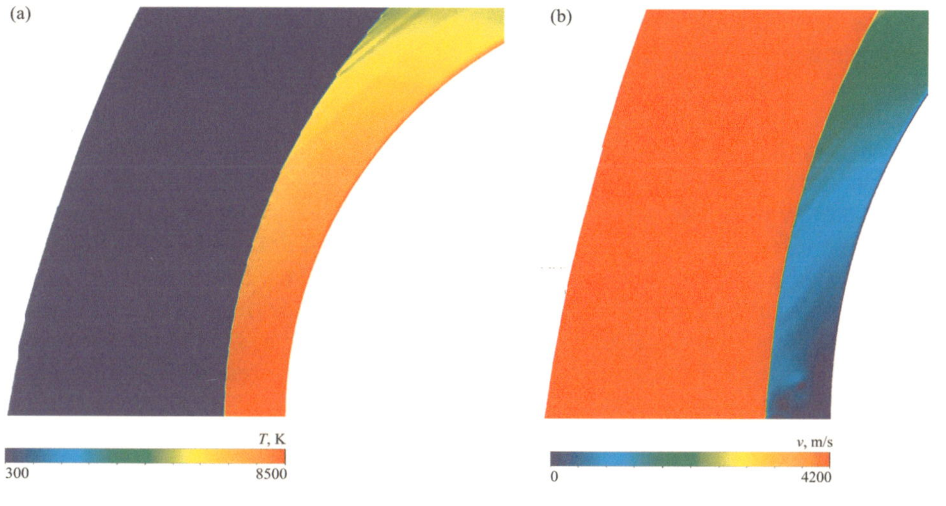

Figure 9.
Temperature contours computed with the Roe scheme and viscous model (a) and velocity contours computed with the AUSM scheme and viscous model (b).

Figures from 9 to 11 show only the part of the domain near the surface of the sphere where a significant impact of the computational options (numerical scheme) on distributions of fluid quantities is observed.

The contours of flow quantities computed with the Roe scheme and AUSM scheme are shown in **Figure 9**. The results obtained show the presence of weak non-physical curvature of the head shock when using the Roe scheme. When using the AUSM scheme, there are no non-physical shock wave distortions, but vortex structures are present near the symmetry axis.

Accounting for viscosity suppresses non-physical perturbations of the numerical solution. With an increase in air temperature, its viscosity increases, which is taken into account using the Sutherland formula (the same dependence is used to take into account the dependence of air thermal conductivity on temperature). Accounting for

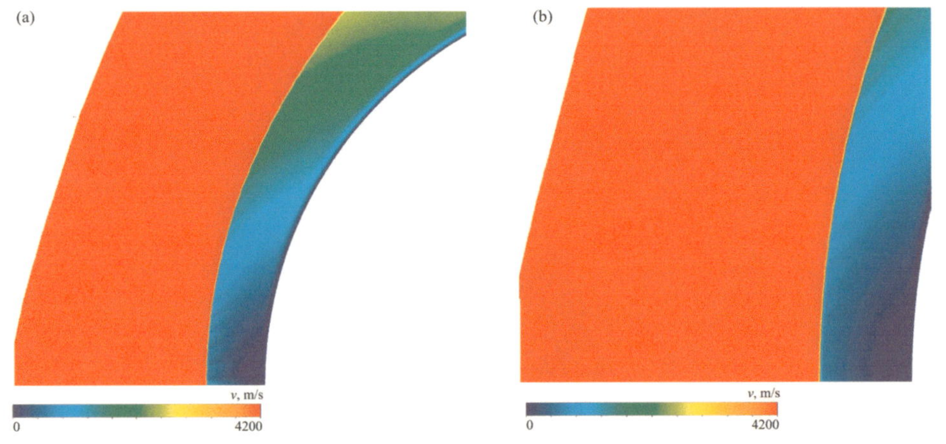

Figure 10.
Velocity contours computed with the Roe scheme (a) and AUSM scheme (b) using variable viscosity.

the change in viscosity with temperature changes reduces the non-physical oscillations. The results of calculations obtained in a laminar formulation, taking into account temperature-dependent viscosity and thermal conductivity of air, are shown in **Figure 10** for Roe and AUSM schemes. The calculation results obtained using the Roe scheme show that the shape of the shock wave has no anomalies, but behind it, there are irregularities in the field of the velocity modulus. The use of the AUSM scheme leads to the fact that there are no vortices near the symmetry axis, but there are also velocity irregularities behind the shock wave.

To account for turbulence, the Reynolds-averaged Navier–Stokes (RANS) equations with the Spalart-Allmares turbulence model are used. The calculation results obtained in the turbulent setting are shown in **Figure 11** for the Roe and AUSM schemes. The shock wave calculated using the Roe scheme has no anomalies and no unevenness of the velocity field. AUSM scheme also leads to results characterized by a smooth shape of the shock wave front.

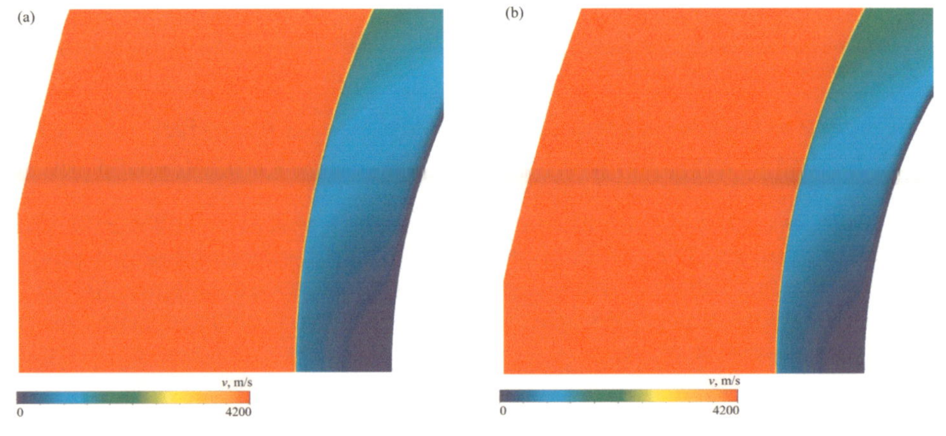

Figure 11.
Velocity contours computed with the Roe scheme (a) and AUSM scheme (b) using the turbulence model.

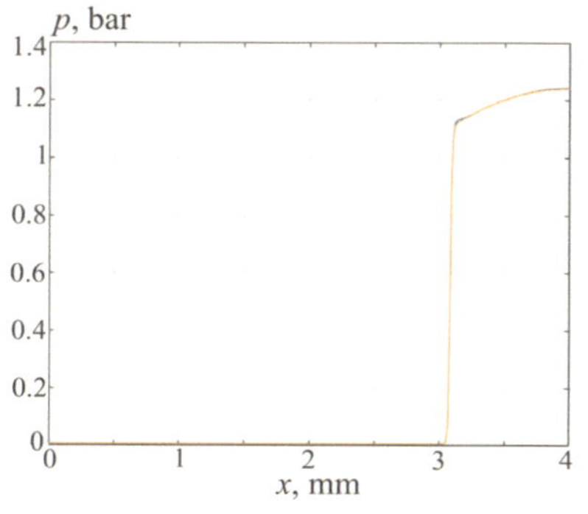

Figure 12.
Centerline pressure distribution computed with the AUSM scheme using a turbulence model.

Figure 13.
Temperature contours computed with equilibrium chemical reactions in high-temperature air.

The pressure distribution along the symmetry axis in **Figure 12** shows that there are no oscillations of the numerical solution near the shock wave. Solutions calculated with Roe and AUSM schemes practically coincide.

Figure 13 displays the results computed with a Mach number of 12 and a high-temperature air model, as obtained using the AUSM scheme. Because of the considerably lower air temperature in this scenario compared to a perfect gas model, there is less compaction at the point where the sphere's surface meets the shock.

The pressure distribution along the centerline, presented in **Figure 14**, presents the position of the shock wave. It differs from that obtained with the perfect gas model (**Figure 12**).

The relation [23] is used to determine the relative thickness of the shock layer in a perfect gas.

$$\frac{\Delta}{D} = K\varepsilon, \varepsilon = \frac{\gamma - 1}{\gamma + 1}\left(1 + \frac{2}{\gamma - 1}\frac{1}{M_\infty^2}\right). \tag{24}$$

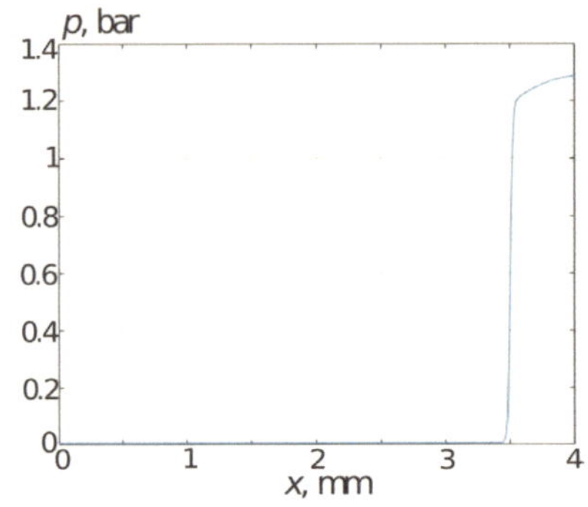

Figure 14.
Centerline pressure distribution computed with the AUSM scheme using equilibrium chemical reactions.

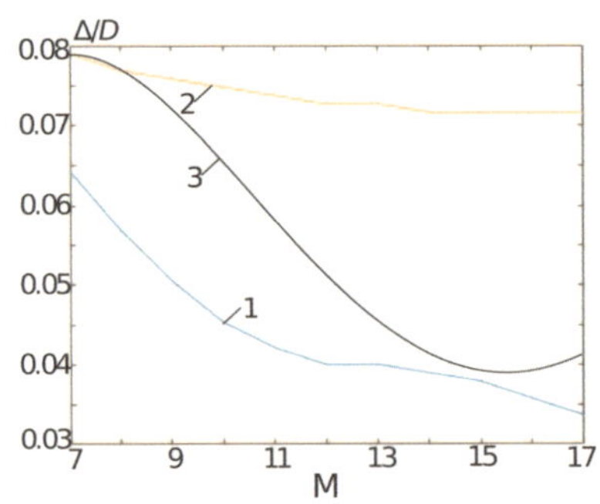

Figure 15.
Dependence of the dimensionless thickness of the shock layer on Mach number.

The proportionality coefficient is denoted by $K = 0.39 - 0.41$, and the parameter ε is the ratio of densities in the shock layer and undisturbed flow ($\varepsilon = \rho_\infty/\rho_s$). A series of calculations is carried out for various flow incident Mach numbers on the sphere, and the outcomes are compared with the experimental data. The dependence of the dimensionless thickness of the shock layer on Mach number is depicted in **Figure 15**. Line 1 represents the dependence of dimensionless distance to the shock wave on Mach number, assuming perfect equilibrium chemical reactions. Line 2 shows the calculated results using a non-reacting constant adiabatic index ($\gamma = 1.4$) perfect gas model. Line 3 shows interpolated experimental data from Ref. [19]. Airflow is considered frozen at low velocity, and the differences between physical experiments and computational results are caused by non-equilibrium effects in air [19]. The experimental data (line 3) at $M = 14$ is most closely approximated by line 1, which approaches them at higher Mach numbers of the approaching flow.

6. Conclusion

Supersonic flow in a stepped channel is numerically simulated. The resultant flow shock-wave structure agrees with the available numerical solutions. The step-reflected shock wave front is smeared into two or three mesh cells. The numerical simulation results show that the solution does not exhibit non-physical oscillations, which are typical of TVD-type and low-order circuits. Even at low Courant numbers, there are noticeable errors in the numerical solution of a hypersonic inviscid airflow around a sphere due to the Roe scheme. The most accurate scheme is the Rusanov scheme.

The method for calculating supersonic high-temperature airflows is integrated into the Ansys Fluent software package using a programming user interface. The capabilities of the standard version of the commercial package have been expanded, which make it possible to use the Ansys Fluent package for modeling supersonic and hypersonic airflows, taking into account high-temperature effects.

Author details

Konstantin Volkov
Kingston University, London, United Kingdom

*Address all correspondence to: k.volkov@kingston.ac.uk

References

[1] Volkov KN, YuN D, Emelaynov VN, Kozelkov AS, Karpenko AG, Teterina IV. Speed-Up of Gas Dynamics Calculations on Unstructured Meshes. Moscow: Publishing House of Physical and Mathematical Literature; 2013. p. 536

[2] Lipanov AM, Karskanov SA. Application of high-order approximation schemes in modeling the deceleration of supersonic flows in rectangular channels. Computational Continuum Mehanics. 2013;**6**:292-299

[3] Zabarko DA, Kotenev VP. Numerical study of laminar flows of a viscous chemically reacting gas near blunt bodies. Vestnik MGTU. Naural Sciences. 2006;**1**:77-95

[4] Kraiko AN, Makarov VE. Explicit analytical formulas determining the equilibrium composition and thermodynamic functions of air for temperatures from 200 to 20000 K. Thermophysics of High Temperatures. 1996;**34**:208-219

[5] Dobrov Y, Karpenko A, Malkovsky S, Sorokin A, Volkov K. Simulation of high-temperature flowfield around hypersonic waverider using graphics processor units. Acta Astronautica. 2023; **204**:745-760

[6] Kulikovsky AG, Pogorelov NV, Semenov AY. Mathematical Questions of the Numerical Solution of Hyperbolic Systems of Equations. Moscow: Publishing House of Physical and Mathematical Literature; 2001. p. 312

[7] Roe PL. Approximate Riemann solvers, parameter vectors, and difference schemes. Journal of Computational Physics. 1981;**43**: 357-372

[8] Rusanov VV. Calculation of the interaction of non-stationary shock waves with obstacles. Computational Mathematics and Mathematical Physics. 1961;**1**:267-279

[9] Liou M-S. A sequel to AUSM. II. AUSM+-up for all speeds. Journal of Computational Physics. 2006;**214**: 137-170

[10] Bulat PV, Volkov KN. WENO schemes for solution of unsteady one-dimensional gas dynamics test problems. Scientific and Technical Journal of Information Technologies, Mechanics and Optics. 2016;**15**:174-180

[11] Bulat PV, Volobuev IA, Volkov KN, Pronin VA. Numerical simulation of regular and Mach reflection of shock wave from the wall. Scientific and Technical Journal of Information Technologies, Mechanics and Optics. 2017;**17**:920-928

[12] Woodward PR, Colella P. The numerical simulation of two-dimensional fluid flow with strong shocks. Journal of Computational Physics. 1984;**54**:115-173

[13] MacCormack RW. Carbuncle computational fluid dynamics problem for blunt-body flows. Journal of Aerospace Information Systems. 2013; **10**:229-239

[14] Isaev SA, Lysenko DA. Testing the fluent package for calculating supersonic flow in a stepped channel. Journal of Engineering Physics and Thermophysics. 2004;**77**:164-167

[15] Isaev SA, Lysenko DA. Testing of numerical methods, convective schemes, flow approximation algorithms and grid

structures on the example of supersonic flow in a stepped channel using CFX and fluent packages. Journal of Engineering Physics and Thermophysics. 2009;**82**: 326-330

[16] Berthon C. Robustness of MUSCL schemes for 2D unstructured meshes. Journal of Computational Physics. 2006; **218**:495-509

[17] Christov I, Popov B. New nonoscillatory central schemes on unstructured triangulations for hyperbolic systems of conservation laws. Journal of Computational Physics. 2008; **227**:5736-5757

[18] Holt M, Hoffman GH. Calculation of hypersonic flow past sphere and ellipsoids. American Rocket Society. 1961;**31**:61-209-1903

[19] Lobb RK. Experimental Measurement of Shock Detachment Distance on Spheres Fired in Air at Hypervelocities/High Temperature Aspects of Hypersonic Flow. Oxford: Pergmon Press; 1964. pp. 519-527

[20] Zhluktov SV, Smekhov GD, Tirsky GA. Rotational-vibrational-dissociation interaction in a multicomponent nonequilibrium viscous shock layer. Fluid Dynamics. 1994;**6**: 166-179

[21] Dobrov Y, Gimadiev V, Karpenko A, Volkov K. Numerical simulation of hypersonic flow with non-equilibrium chemical reactions around sphere. Acta Astronautica. 2022;**194**:468-479

[22] Emelyanov VN, Karpenko AG, Volkov KN. Simulation of hypersonic flows with equilibrium chemical reactions on graphics processor units. Acta Astronautica. 2019;**163**:259-271

[23] Golovachev YP. Numerical Modeling of Viscous Gas Flows in the Shock Layer. Moscow: Nauka; 1996. p. 376

Chapter 5

Evolution of Hall Thruster Technology toward Future Space Exploration

Ikkoh Funaki, Hiroki Watanabe and Yuya Oshio

Abstract

This paper provides a brief overview of Hall thruster R&Ds toward exploration worldwide and efforts for a highly fuel-efficient Hall thruster. The development goal is to expand the Hall thruster technology for all-electric propulsion geostationary satellites that operate at a few operational points toward high specific impulse and wide power ranges. These characteristics are important, particularly for space exploration. Until now, laboratory model thrusters and power processing units have been fabricated and tested using 6-kW-class thrusters; also, extremely high-voltage Hall thrusters above 3000-second specific impulse are experimentally investigated.

Keywords: electric propulsion for spacecraft, Hall thruster, all-electric propulsion satellite, space exploration, high specific impulse

1. Introduction

Hall thruster is a type of electric propulsion that provides fuel-efficient spacecraft maneuvers for various space missions. As the schematic illustration in **Figure 1** shows, the Hall thruster features an annular channel for plasma production and acceleration. When electrons from a hollow cathode are provided into the channel, the Hall current in the azimuthal direction is induced as a cross-product of both the magnetic field and the electrostatic field; both are exerted via coil magnets or the anode-to-cathode circuit. When propellant (usually xenon (Xe) due to its ease of ionizing) flows into the channel, electron-neutral collisions happen, and a high-density plasma is produced. Then, the plasma is accelerated to form a jet via the Lorentz force. The thrust force produced is expressed as

$$F = \dot{m} u_{jet} \quad (1)$$

where $\dot{m}$ is the mass flow rate and u_{jet} is the exhaust velocity of the plasma jet. Usually, fuel efficiency is expressed by specific impulse I_{sp} as,

$$I_{sp} = \frac{F}{\dot{m} g} = \frac{u_{jet}}{g} \quad (2)$$

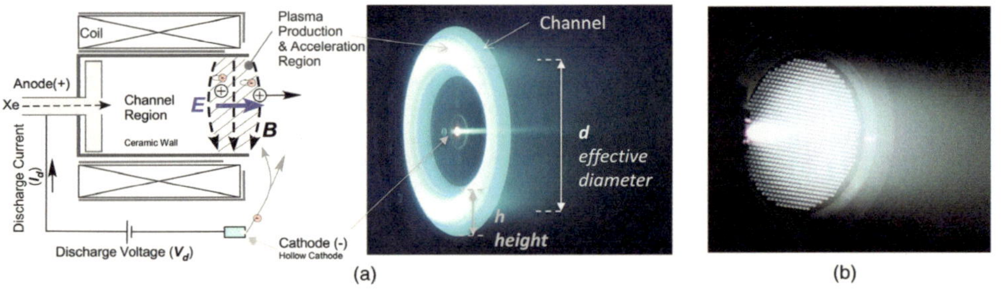

Figure 1.
Hall thruster and ion thruster: the pictures show Japan Aerospace Exploration Agency's (JAXA's) 6-kW Hall thruster for Engineering Test Satellite-9 (ETS-9) and Ion thruster for the Hayabusa Series. Hall thrusters feature higher density plasma jets than gridded ion thrusters, whose ion current density is limited by the space charge effects; hence, compact thruster design becomes possible for Hall thrusters. (a) 6 kW class Hall thruster for ETS-9 and its schematics. (b) Ion thruster for the Hayabusa series.

where I_{sp} means how many seconds a kilogram of propellant produces a kilogram force of thrust.

Due to sufficiently high I_{sp} above 1500 s and compact, hence lightweight system against ion thrusters, Hall thrusters are widely used for current space missions, particularly for Earth-orbiting missions, such as Geostationary orbit (GEO) and low-earth orbit (LEO) satellites, and constellations. Most of them operate in a single or narrow operational range, but in terms of I_{sp}, the Hall thrusters are inferior to the ion thrusters. For space exploration, a highly fuel-efficient Hall thruster as I_{sp} > 3000 s must endure a long-term operation. Hall thrusters for exploration are thus necessary to widen operational ranges compared to those of the existing thrusters, particularly in the higher I_{sp} range. This chapter focuses on high I_{sp} Hall thrusters for exploration, introducing the research and development (R&D) status in the world and activities at JAXA.

2. The trend of R&Ds toward exploration

Hall thrusters have expanded worldwide since all-electric propulsion (all-EP) satellites became apparent in the market [1]. These all-EP satellites usually have Hall thrusters as their main propulsion, enabling a cost-effective payload transfer to a target orbit. In typical GEO missions, after a launch into a geostationary transfer orbit, all the maneuvers are provided by EP systems. With this method, the payload efficiency of the all-EP satellite is twice as high as that of a conventional satellite using chemical propulsion. The number of satellites employing all-EP systems is therefore increasing, and most satellite developers have their lineups of all-EP satellites so that customers can select a desired satellite bus to provide their services. The technology of the all-EP system has been expanded for various uses, as explained in **Figure 2**. The first generation of the all-EP satellites provides maneuverability for several-ton class satellites, enabling both a high-thrust orbital transfer and a low-thrust orbital-and-attitude correction. This technology can be downsized for constellations that typically weigh around 100 kg, but when a small delta-V (ΔV) is satisfactory, single-mode operation at a lower thrust efficiency will be sufficient. Under such a situation, in addition to high-performance xenon Hall thrusters, low-performance thrusters with various propellants will be applicable. For micro- or nano-scale satellites (weighing 10 kg or less), despite simply downsizing Hall thrusters for bigger satellites, various

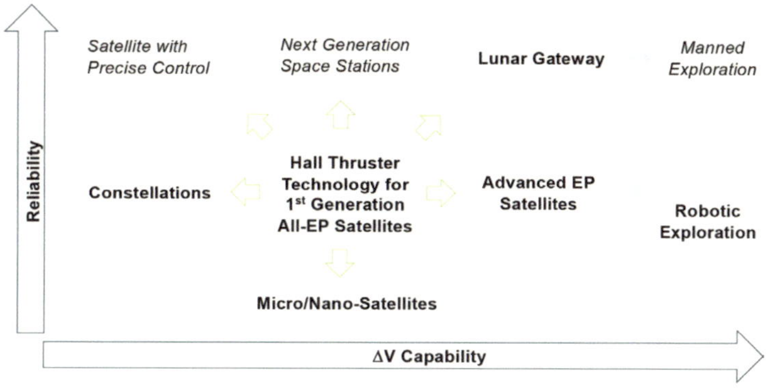

Figure 2.
Trend of current and future EP technology.

thruster options are pursued that simplify the architecture and drastically lower the cost against the existing thrusters [2, 3]. The thruster market of smaller spacecraft is, hence, rich in diversity.

The opposite direction for R&D is toward higher ΔV capability and high reliability, as the upper and right directions in the figure target human systems

Mission	Destination	Launch year	EP system	Power and I_{sp}	Status and source
Deep Space 1	Asteroid 9969 Braille	1998	NSTAR (Ion Thruster)	2 kW 3100 s	Completed [4]
SMART-1	Moon	2003	PPS 1350 (Hall Thruster)	1.2 kW 1600 s	Completed [5]
Hayabusa	Asteroid Itokawa	2003	μ 10 (Ion Thruster)	0.4 kW 3000 s	Completed [6]
Dawn	Asteroid Vesta, Ceres	2007	NSTAR (Ion Thruster)	2.5 kW 3100 s	Completed [7]
Hayabusa2	Asteroid Ryugu	2014	μ 10 (Ion Thruster)	0.4 kW 2900 s	In progress [8]
PROCYON	Asteroid	2014	I-COUPS (Ion Thruster)	0.04 kW N/A	Abandoned [9]
BepiColombo	Mercury	2018	T6 (Ion Thruster)	4.6 kW 4200 s	In progress [10] [11]
DART	Asteroid Dimorphos	2021	NEXT-C (Ion Thruster)	7.4 kW 4200 s	Completed [12]
Psyche	Asteroid Psyche	2023	SPT-140 (Hall Thruster)	4.5 kW 1800 s	In progress [13]
PPE	Lunar Gateway	2026	Hermes/BHT-6000 (Hall Thrusters)	12 kW 2600 s 6 kW 2600 s	In preparation [14]
DESTINY+	Asteroid Phaethon	2026	μ 10 (Ion Thruster)	0.4 kW 3000 s	In preparation [15]
MSR-ERO	Earth-Mars round trip	TBD	RIT-2X (Ion Thruster)	7.5 kW 3800 s	In preparation [16]

Table 1.
EP systems for exploration.

or robotic explorations. In this direction, high-performance and high-reliability systems are necessary compared to the first-generation all-EP satellites. Regarding electric propulsion for space exploration, selected missions are described with EP systems employed in **Table 1**. The table shows that most missions have employed ion thrusters instead of Hall thrusters, primarily due to their high I_{sp} necessary for deep space missions. Some ion thrusters provided a high I_{sp} of about 4000 s, and the values of I_{sp} of Hall thrusters so far were not so high. However, there are some cases where the Hall thruster was used. In the past, SMART-1 (Small Missions for Advanced Research in Technology-1) to the Moon, and recently, an asteroid mission, Psyche, employed Hall thrusters.

The Psyche spacecraft employed Hall thruster technology based on the commercial all-EP satellite bus [13], but future exploration missions require a much quicker and more efficient transfer by optimizing the thruster for each exploration mission. In terms of these requirements, including lifetime for long-term operations, the advanced electric thruster system (AEPS) of the Power and Propulsion Element (PPE) for the Lunar gateway will provide the most advanced flight thrusters in the world (HERMeS (Hall Effect Rocket with Magnetic Shielding) and BHT-6000) that can provide a high I_{sp} of 2600 s at a high discharge voltage of 600 V with deep throttling capability [17]. The thrusters also employed magnetic shielding technology, which realizes very low erosion rates of the discharge channel of Hall thrusters [18].

Future missions require higher I_{sp} as exploration regions evolve deep into the solar system, and Hall thrusters at higher I_{sp} will contribute to future explorations. Following such objectives, there were so many studies to try a higher I_{sp} Hall thruster, as given in **Table 2**. Like in the cases of SPTs and BHTs, up to the early 2020s, trials were conducted for Hall thrusters of 10-cm diameter or nearby sizing. These researchers indicated the possibility of operation above 3000 s. The upper limit was 1200 V and 3700 s, and very high I_{sp} was successfully obtained with the thrusters in a high-power density regime. Later, along with several efforts to obtain a higher I_{sp}, R&Ds by National Aeronautics and Space Administration Jet Propulsion Laboratory (NASA JPL) and Glenn led to the high-I_{sp} Hall thruster developments, clarifying the scaling law for high-voltage Hall thrusters, and the long-life thruster technology

Thruster	d [mm]	Discharge voltage [V]	Power [kW] and I_{sp} [s]	Year and source
SPT-80	80	850	2.5 kW 3200 s	1998 [19]
SPT-1	100	1250	2.5 kW 3200 s	2001 [20]
BHT-1000	N/A	1000	2.4 kW 3200 s	2001 [21]
BHT 1500	N/A	1200	2.8 kW 3400 s (anode)	2005 [22]
NASA-173 M	N/A	1000	9.6 kW 3400 s	2004 [23]
SPT-115	115	1100	3.3 kW 3700 s	2006 [24]
SPT-140	140	1000	4 kW 3600 s	2006 [24]
NASA-HERMeS	N/A	700	12 kW 3000 s	2016 [17]
HVHT-100	100	900	3.5 kW 3200 s	2020 [25]
NASA-H10	N/A	800	10 kW 3400 s	2024 [26]

Table 2.
Selected R&Ds of high I_{sp} *Hall thrusters.*

by the magnetic shielding method was established [18]. The current R&D target by NASA is high-power Hall thrusters for future Mars missions with a higher I_{sp} with the magnetic shielding technology.

3. Pursuit of extremely high-voltage Hall thruster

From the list mentioned in **Table 2**, the values of I_{sp} up to 3700 s were obtained. It is the fundamental question of whether a higher I_{sp} is possible and what is the highest possible I_{sp} by Hall thrusters. To identify and possibly exceed the current limit, the authors work on laboratory experiments with small Hall thrusters [27, 28]. After surveying several channel geometries, we found that the influence of channel geometries was significant for extremely high-voltage operation, and I_{sp} up to 4000 was obtained as the temporary maximum, as is described in the following.

To increase I_{sp}, the discharge voltage (V_d) needs to be increased. Based on the theory of Hall thrusters, I_{sp} is related to V_d as

$$I_{sp} = \frac{\eta_u}{g}\sqrt{\frac{2eV_d}{m_i}} \tag{3}$$

where η_u is the propellant utilization efficiency, e is the elementary charge, and m_i is the mass of an ion. Following the above dependency, there is a possibility that quite a high I_{SP} can be obtained at an extremely high discharge voltage. Among various configurations, two thruster arrangements in **Figure 3** were selected and described in this chapter. One is the h16 thruster, whose channel height (h) is 16 mm with the effective channel diameter, d, of 60 mm, corresponding to the typical channel aspect ratio as a Hall thruster. The configuration of another thruster (h8 thruster) is h = 8 mm and d = 118 mm, corresponding to a narrower channel expecting a more intensive plasma-to-channel reaction than ordinary Hall thrusters. A comparison is made to find out how their discharge characteristics depend on the geometries.

Figure 4 shows the experimental results of both thrusters. In **Figure 4(a-1)**, **(a-2)**, thrust values (F) are plotted against the maximum radial magnetic field B_{r_max} for V_d. In **Figure 4(a-1)**, the data are limited to three discharge voltages, V_d = 300,

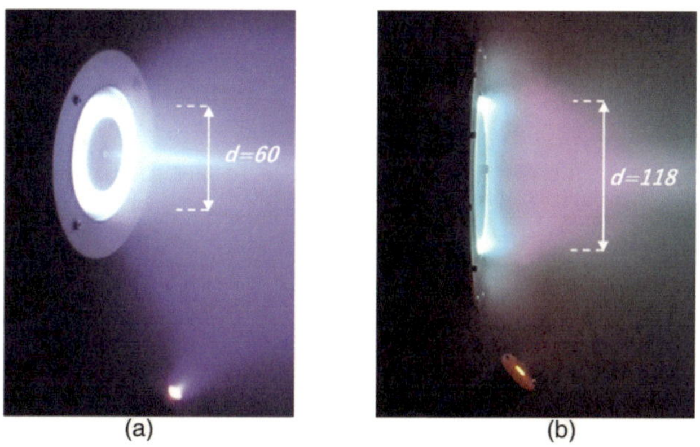

(a) (b)

Figure 3.
Small xenon Hall thrusters for high discharge voltage experiment. (a) h16 channel (d = 60 mm, h = 16 mm). (b) h8 channel (d = 118 mm, h = 8 mm).

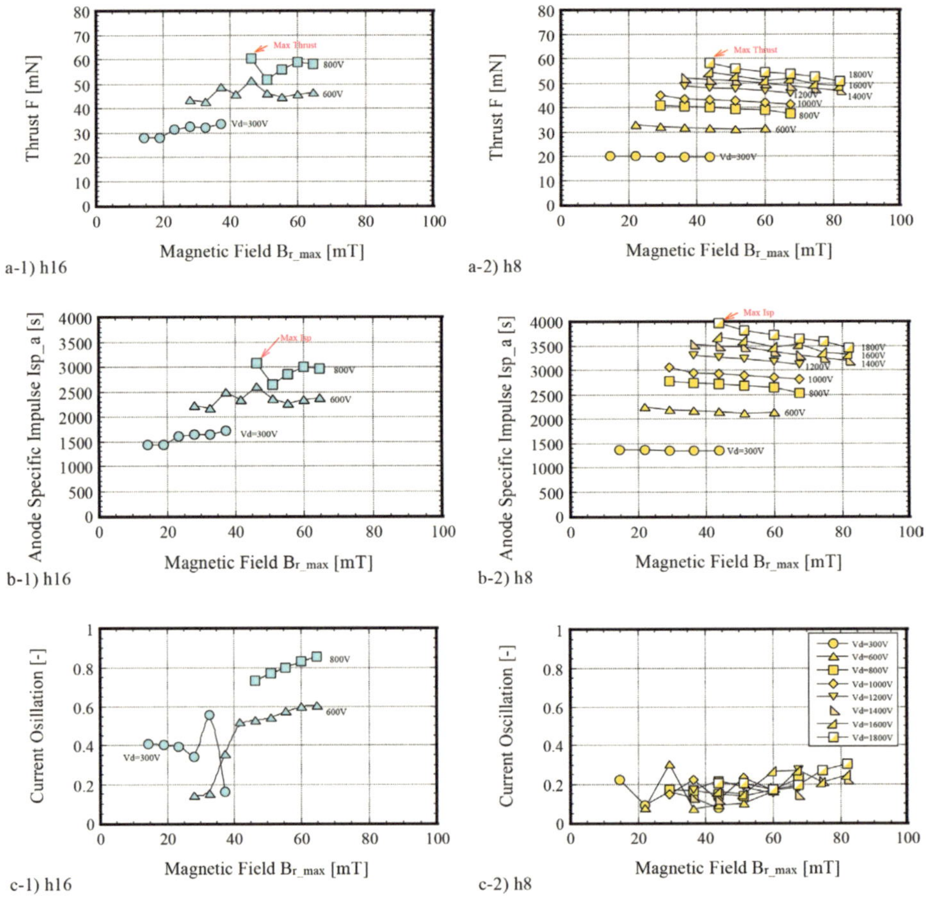

Figure 4.
Characteristics of small xenon Hall thrusters; (a) Thrust vs. Br_max, *(b) anode* Isp *vs.* Br_max, *and (c) discharge current oscillation vs.* Br_max. *Xenon flow rates were 2 mg/s for h16 thruster and 1.5 mg/s for h8 thruster, respectively.*

600, and 800 V. The highest thrust, F = 60 mN, was obtained for V_d = 800 V and B_{r_max} = 45 mT. The corresponding anode I_{sp} for this condition is 3080 s, as highlighted in **Figure 4(b-1)**. The data in the figures start from a certain B_{r_max} value. For example, in **Figure 4(a-2)**, the h8 thruster at 1800 V starts from the point B_{r_max} = 44 mT, below which discharge could not continue. This is related to the scaling relation of Hall thrusters, that is, the ratio of the radius of electron Larmor motions r_{Le} to the discharge length, L,

$$\frac{r_{Le}}{L} = \frac{m_e v_e}{eBL} \ll 1 \quad (4)$$

where m_e is the mass of the electron, v_e is the electron velocity, and e is the elementary charge. Eq. (4) corresponds to the fact that when V_d is high, v_e also becomes high, and a large magnetic field B is necessary to maintain a small r_{Le}/L value. The corresponding current oscillation is shown in **Figure 4(c-1)**. The oscillation was evaluated by

$$\overline{I_d} = \frac{1}{\tau}\int_0^\tau I_d dt \tag{5}$$

and

$$\Delta I_d = \frac{1}{\overline{I_d}}\sqrt{\frac{\int_0^\tau \left(\overline{I_d} - I_d\right)^2 dt}{\tau}} \tag{6}$$

At 600 V, the amplitudes of discharge current oscillation were 15–60% of the mean current value, and at 800 V, the amplitudes were around 80%. These are very high oscillations. It was found that discharge was not stable beyond 800 V due to excessive oscillation in the field, and hence, high I_{sp} operation is suspended by the increased current oscillation in this h16 configuration. The h8 configuration shows different results and tendencies. As in **Figure 4(a-2)**, thrust values are increased up to 1800 V, and operation is possible for much higher V_d for h8 with low current oscillation (**Figure 4(c-2)**). According to Eq. (3), it is expected that thrust (and I_{sp}) is proportional to $\sqrt{V_d}$, and this trend appeared in the figure. The maximum thrust of the h8 thruster was 58 mV at 45 mT and 800 V, corresponding to 3970 s, as plotted in **Figure 4(b-2)**. Smaller F and higher I_{sp} result from smaller $\dot{m}$ = 1.5 mg/s in this h8 case. To summarize the comparison between two configurations, anode I_{sp} and anode efficiencies are plotted against V_d in **Figure 5** for the maximum anode efficiency cases. It is observed that the h8 thruster can work at a much higher voltage, and due to the wide operational range, a much higher I_{sp} was possible.

In the following, discussions are made to explain the discrepancy between h8 and h16 thrusters. In the case of the h16 thruster, a plasma distinction was encountered at 1000 V. Looking at the data at 800 V, high current oscillation above 70% was found. When such highly oscillatory behavior is encountered, the channel flow is considered nearly fully ionized with Xe^+, and further ionization is hindered due to the energy gap between Xe^+ and Xe^{2+}. A thruster with a moderate channel width, i.e., the h16 thruster, seems to confront this limit. In the h8 thruster case, however, I_{sp} continuously increases as V_d increases up to 1800 V. It is anticipated that discharge loss is significant in this regime, which is explained by the lower anode efficiencies than the case for h16, as plotted in **Figure 5(b)**. In this case, the current oscillation was not so intensive but stable throughout the region, indicating that the ionization ratio was not so high; this means that the electron energy stays at a moderate value, either by frequent collision with the channel wall or by collisions with neutrals. It can be

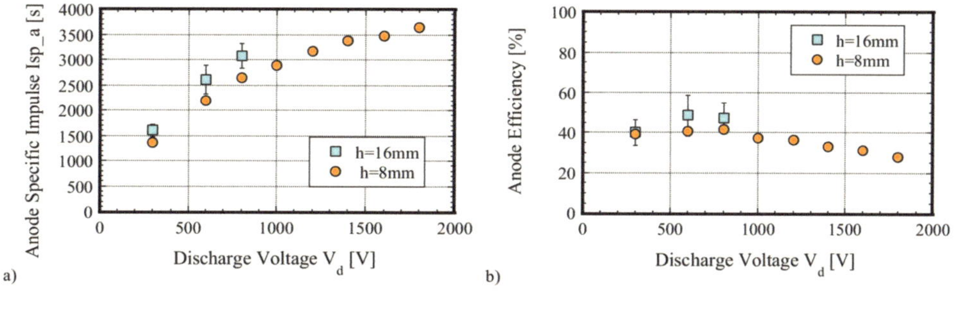

Figure 5.
Performance of small Hall thrusters: (a) Anode Isp *vs.* V_d*, and (b) anode efficiency vs.* V_d*. For each* V_d*, conditions for the maximum anode efficiency were selected and plotted.*

said that high I_{sp} is obtained at the expense of low-thrust performance. With such a moderate v_e value, the h8 thruster keeps another scaling parameter, the Hall parameter, constant:

$$\Omega_e = \frac{\omega_{ce}}{v_{en}} = \frac{eB}{m_e n_n \sigma_{en} v_e} \tag{7}$$

With a moderately high B value a moderately high neutral density, n_n, or equivalently, high-power density seems suitable for high I_{sp} operation. By surveying this high I_{sp} regime, performance improvement is still being tried in the authors' laboratory.

In summary, in the above experiment, I_{sp} reached as much as 3970 s by balancing the loss, ionization, and acceleration [28]. This means further tuning the balance between them it may be possible to obtain improved thrust efficiency by optimizing the small Hall thruster. For extremely high-voltage operations, however, the lifetime issue is still unsolved, so reducing the loss to the channel wall is necessary.

4. Pursuit of medium-power high-voltage Hall thruster system

Although the extremely high-voltage Hall thruster research shown stays in a preliminary stage, step-by-step implementation of high I_{sp} capability to practical thruster systems is prepared as introduced in this section starting from moderately high I_{sp} below 3000 s. JAXA has been working on medium-power Hall thruster development since 2015, and the R&D program includes several developments, including the flight demonstration onboard the Engineering Test Satellite-9 (ETS-9) [29]. Also, for future evolution of the Hall thruster lineup, a laboratory system with a power processing unit was arranged and tested to pursue higher I_{sp} operation of a full-scale 6-kW Hall thruster [30]. Only a moderately high voltage is tested to get an initial design of the overall system. **Figure 6** depicts the laboratory model thruster's performance with the power processing unit laboratory model; in the figure, a few illustrations of configuring variable voltage with specified DC/DC converters were also plotted [31, 32]. The maximum voltage (730 V in the 6-kW case) is obtained when the converters are connected in series. Vice versa, when the converters are connected in parallel, the most significant current (40 A in the 6-kW case) is obtained at a discharge voltage of 200 V (the lowest). Likewise, stepwise voltages of 200, 400, 600, and 800 V are available by an appropriate choice of converter connection. **Figure 6(c)** provides thrust efficiency, and **Figure 6(d)** indicates that a high thrust-to-power (T/P) ratio of as much as 70 mN/kW is possible at a lower I_{sp}.

At a higher I_{sp}, 62 mN/kW was obtained for the intermediate I_{sp} of 2000 s, and the highest I_{sp} was 2750s but with a smaller T/P value of 40 mN/kW. Although they did not reach the expected values of 80 mN/kW at 1300 s and 40 mN/kW at 3000 s, further optimization is going on, but the performance shows the possibility of a wide-range mode using a single thruster head and the variable power processing unit design. In the exploration application, quick acceleration after a launch will be needed, but a more efficient transfer is preferable during an orbital transfer in a power-decreasing environment. This means both high thrust and high I_{sp} modes are necessary and selectively used to optimize the transfer. Following this successful wide-range test, a system with a wide I_{sp} range can further optimize the mission.

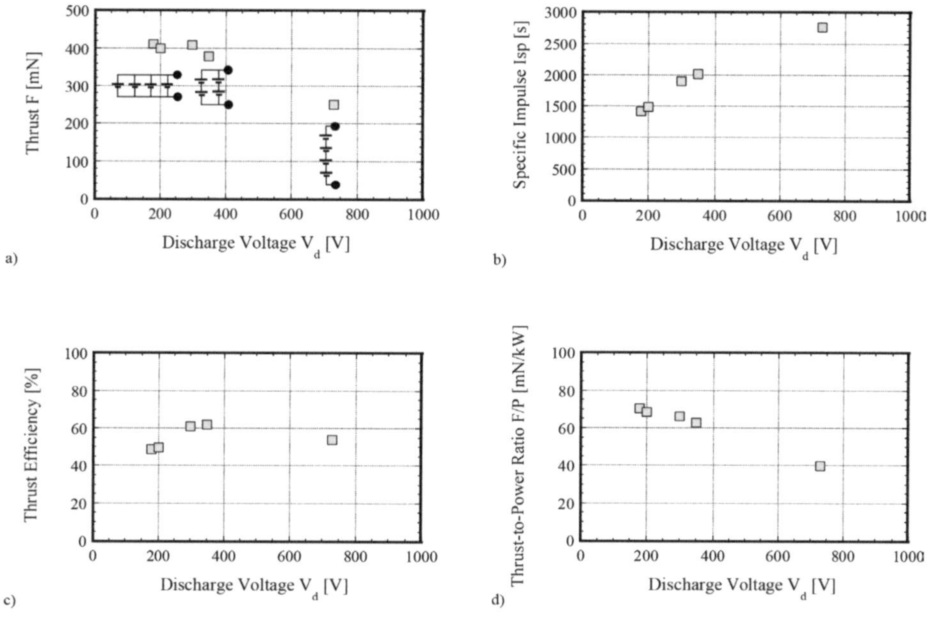

Figure 6.
Initial performance of a medium-power xenon Hall thruster breadboard model (d = 150 mm and 6-kW operation); (a) Thrust, (b) I_{sp}*, (c) thrust efficiency, and (d) thrust-to-power ratio are plotted against* V_d.

5. Summary

The current status and future prospects of the application of Hall thruster technology toward exploration are discussed in this chapter. In addition to ion thrusters, exploration missions with Hall thrusters have appeared and are expected to grow. To overcome the limited I_{sp} as exploration missions, intensive studies were conducted to operate Hall thrusters at extremely high I_{sp} above 3000 s. In addition to the high-power and high I_{sp} missions, such as NASA's human exploration program, there is a need for small and high I_{sp} Hall thrusters. Results of some experimental trials were provided, and there was an operational limit in terms of I_{sp} of Hall thruster due to the nearly full ionization condition inside the discharge channel. The limit stays between 3000s to 4000s depending on the channel configuration, which shows the possibility of controlling the electron energy for a high I_{sp} Hall thruster by changing the discharge channel. Considering the issues associated with lifetime at higher I_{sp} and limited power supply systems, a thruster system for high I_{sp} thrusters should be further investigated. In future exploration missions, rapid and efficient transfer is desired by enabling high I_{sp} and high-thrust thrusters.

Acknowledgements

Part of this work is supported by JAXA and the Japan Society for the Promotion of Science KAKENHI grant number 20H05656.

Author details

Ikkoh Funaki[1*], Hiroki Watanabe[1] and Yuya Oshio[2]

1 Japan Aerospace Exploration Agency (JAXA), Sagamihara, Japan

2 Ryukoku University, Otsu, Japan

*Address all correspondence to: funaki@isas.jaxa.jp

References

[1] Lev DR, Emsellem GD, Hallock AK. The rise of the electric age for satellite propulsion. New Space. 2017;**5**:4-14. DOI: 10.1089/space.2016.0020; 5. Epub ahead of print 2017

[2] Lev D, Myers RM, Lemmer KM, et al. The technological and commercial expansion of electric propulsion. Acta Astronautica. 2019;**159**:213-227. DOI: 10.1016/j.actaastro.2019.03.058. Epub ahead of print 2019

[3] O'reilly D, Herdrich G, Kavanagh DF. Electric propulsion methods for small satellites: A review. Aerospace. 2021;**8**: 1-30. DOI: 10.3390/aerospace8010022. Epub ahead of print 2021

[4] Polk JE, Brinza D, Kakuda RY, et al. Demonstration of the NSTAR ion propulsion system on the deep space one mission. 27th International Electric Propulsion Conference, IEPC-01-75. Pasadena, 2001.

[5] Koppel CR, Estublier D. SMART-1 primary electric propulsion sub-system the flight model. In: 28th International Electric Propulsion Conference, IEPC-03-0205. Toulouse. 2003

[6] Kuninaka H, Nishiyama K, Funaki I, et al. Powered flight of electron cyclotron resonance ion engines on Hayabusa explorer. Journal of Propulsion and Power. 2007;**23**:544-551. DOI: 10.2514/1.25434. Epub ahead of print 2007

[7] Garner CE, Rayman MD, Brophy JR, et al. The dawn of Vesta science. In: 32nd International Electric Propulsion Conference, IEPC-2011-326. Wiesbaden. 2011

[8] Nishiyama K, Hosoda S, Ueno K, et al. Development and testing of the HAYABUSA2 ion engine system. Transactions of the Japan Society for Aeronautical and Space Sciences, Aerospace Technology Japan. 2016. Epub ahead of print 2016;**14**:Pb_131-Pb_140. DOI: 10.2322/tastj.14.pb_131

[9] Koizumi H, Kawahara H, Yaginuma K, et al. Initial flight operations of the miniature propulsion system installed on small space probe: PROCYON. Transactions of the Japan Society for Aeronautical and Space Sciences, Aerospace Technology Japan. 2016;**14**:Pb_13-Pb_22. DOI: 10.2322/tastj.14.pb_13. Epub ahead of print 2016

[10] Grubisic AN, Clark S, Wallace N, et al. Qualification of the T6 ion thruster for the BepiColombo mission to the planet mercury. In: 32nd International Electric Propulsion Conference, IEPC-2011-234. Wiesbaden. 2011

[11] Snyder JS, Goebel DM, Hofer RR, et al. Performance evaluation of the T6 ion engine. Journal of Propulsion and Power. 2012;**28**:371-379. DOI: 10.2514/1.B34173. Epub ahead of print 2012

[12] Fisher J, Ferriauolo B, Hertel T, et al. NEXT-C flight ion propulsion system development status. In: 35th International Electric Propulsion Conference, IEPC-2017-218. Atlanta. 2017

[13] Oh DY, Alonge N, Bairstow SH, et al. Launch and initial checkout of the psyche spacecraft for NASA's discovery program. In: 38th International Electric Propulsion Conference, IEPC-2024-126. Toulouse. 2024

[14] Herman DA, Gray T, Johnson I, et al. Development and qualification status of the electric propulsion systems for

the NASA PPE mission and gateway program. In: 37th International Electric Propulsion Conference, IEPC-2022-465. 2022

[15] Tabata K, Tsukizaki R, Imai S, et al. Development status of the microwave ion engine for JAXA's DESTINY+ mission. In: 38th International Electric Propulsion Conference, IEPC-2024-462. Toulouse. 2024

[16] Kirch P, Colin D, Ganteil D, et al. Plasma propulsion subsystem development for MSR-ERO. In: 38th International Electric Propulsion Conference, IEPC-2024-655. Toulouse. 2024

[17] Conversano RW, Hofer RR, Sekerak MJ, et al. Performance comparison of the 12.5 kW HERMeS Hall thruster technology demonstration units. In: 52nd AIAA/SAE/ASEE Joint Propulsion Conference, 2016. 2016. DOI: 10.2514/6.2016-4827. Epub ahead of print 2016

[18] Hofer RR, Gallimore AD. The role of magnetic field topography in improving the performance of high-voltage Hall thrusters. In: 38th AIAA/ASME/SAE/ASEE Joint Propulsion Conference and Exhibit. 2002. DOI: 10.2514/6.2002-4111. Epub ahead of print 2002

[19] Kirn V, Kozlov V, Lazurenko A, et al. Development and characterization of small SPT. In: 34th AIAA/ASME/SAE/ASEE Joint Propulsion Conference and Exhibit. 1998. DOI: 10.2514/6.1998-3335 Epub ahead of print 1998

[20] Manzella D, Jacobson D, Jankovsky R. High voltage SPT performance. In: 37th Joint Propulsion Conference and Exhibit. 2001. DOI: 10.2514/6.2001-3774. Epub ahead of print 2001

[21] Pote B, Tedrake R. Performance of a high specific impulse Hall thruster. In: 27th International Electric Propulsion Conference, IEPC-01-35. Pasadena. 2001

[22] Szabo JJ, Azziz Y. Characterization of a high specific impulse xenon Hall effect thruster. In: 29th International Electric Propulsion Conference, IEPC-2005-324. Princeton. 2005. Available from: https://www.researchgate.net/publication/238082917

[23] Hofer RR. Development and characterization of high-efficiency, high-specific impulse xenon Hall thrusters [Ph.D. thesis]. USA: The University of Michigan; 2004. Available from: http://www.sti.nasa.gov

[24] Lazurenko A, Vial V, Bouchoule A, et al. Dual-mode operation of stationary plasma thrusters. Journal of Propulsion and Power. 2006;**22**:38-47. DOI: 10.2514/1.8279

[25] Xia G, Liu X, Li H, et al. Performance improvement of Hall thrusters with high discharge voltage. Vacuum. 2020;**172**:109052, 1-7. DOI: 10.1016/j.vacuum.2019.109052. Epub ahead of print 2020

[26] Hofer RR, Simmonds JB, Goebel DM, et al. The H10 high power density Hall thruster. In: 38th International Electric Propulsion Conference, IEPC-2024-515. Toulouse. 2024

[27] Watanabe H, Matsumoto Y, Amano K, et al. The effect of discharge channel and magnetic field configurations on discharge characteristics of high voltage Hall thrusters. Aerospace Technology Japan, the Japan Society for Aeronautical and Space Sciences (in Japanese); in press

[28] Sawanishi Y, Funaki I, Watanabe H. Design and characterization of a high-voltage Hall thruster. In: 38th

International Electric Propulsion Conference, IEPC-2024-214. Toulouse. 2024

[29] Funaki I, Sano T, Fukatsu T, et al. Development and qualification of a 6-kW-class Hall thruster subsystem for ETS-9. Journal of Evolving Space Activities. 2024;**2**:1-5. Article ID:122. DOI: 10.57350/jesa.122

[30] Funaki I, Sano T, Fukatsu T, et al. The updated R&D status of 6-kW-class Hall-thrusters at JAXA. In: 37th International Electric Propulsion Conference, IEPC-2022-328. Boston. 2022

[31] Matsunaga Y, Takahashi T, Watanabe H, et al. Wide-output range power processing unit for 6-kW Hall thruster. IEEE Transactions on Aerospace and Electronic Systems. 2022;**58**:1609-1620. DOI: 10.1109/TAES.2021.3117520

[32] Matsunaga Y, Watanabe H, Cho S, et al. Development of a wide-output range power processing unit for 6-kW Hall thruster. In: 37th International Electric Propulsion Conference, IEPC-2022-547. Boston. 2022

Chapter 6

Issues of the Satellite Formation Control

Zaure Rakisheva, Nazgul Kaliyeva, Nursultan Doszhan, Anna Sukhenko and Gulama Garip Alisher Ibrayev

Abstract

The problem of satellite formation motion is considered, described by the modified Lawden-Sedwick-Schweighart equations. The formation mission is real-time Earth observation. The primary satellite with a mass of 65 kg moves in a geostationary orbit. Three accompanying spacecrafts, positioned in the plane of the tetrahedron base, are with masses of 50 kg each. A control algorithm has been developed to ensure a sufficient level of formation stability in the presence of a disturbed reference orbit and various uncertainties, based on robust control algorithms such as H2 and H∞. The maximum control force required to maintain the satellite formation's configuration using these control algorithms is 7 mN. It is necessary to determine the types of propulsion systems that meet these requirements. Among the electric propulsion systems available on the market, such as ion engines and Hall thrusters, the specific impulse allows them to generate a maximum thrust of about 15 mN and a minimum of about 6.6 mN. Comparing the obtained thrust values with the control force required to maintain the satellite configuration, it can be concluded that the developed mathematical framework can be successfully implemented in the satellite formation's control system, ensuring the control of satellite motion and their relative positions.

Keywords: satellite formation motion, geostationary orbit, formation stability, control algorithm, propulsion systems

1. Introduction

To effectively monitor large-scale emergencies such as wildfires and floods in real-time, the use of remote sensing satellites in high-altitude orbits is critical. This paper explores the use of a satellite formation in geostationary orbit, aimed at providing extensive area coverage while maintaining high temporal and spatial resolution. The proposed formation operates as a synthetic aperture telescope, modeled after a Fizeau interferometer, which enables spatial resolution comparable to that of a single advanced optical telescope [1]. By leveraging a multi-aperture design, this approach allows the formation to capture high-quality images of the Earth's surface, utilizing three or more satellites as optical elements [2–8].

The satellite formation presented in this study operates in the infrared spectrum and consists of three mirror satellites arranged in a single plane. An imaging satellite is positioned above this plane at a distance that corresponds to the telescope's focal length. Structurally, this satellite formation takes the shape of a tetrahedron or a regular pyramid. To achieve high-resolution monitoring, it is essential to maintain the precise geometric configuration of the formation, requiring a high degree of accuracy in positioning and coordination [9].

In previous research [10], the authors examined various control methods and algorithms for controlling the motion and maintaining the configuration of satellite formations. In this study, we further investigate the implementation of these control algorithms, focusing on the feasibility of using propulsion systems to maintain the formation's configuration in real-world scenarios. We explore how different propulsion technologies can be integrated to support the required precision in satellite positioning, ensuring the stability of the formation over long-duration missions.

2. Mathematical model of the satellite formation

The satellite formation represents a synthetic aperture telescope for forest fire monitoring and consists of four satellites forming a tetrahedron or regular pyramid. The imaging satellite S0 is located above the three mirror satellites S1, S2, and S3, which are located in the plane of the tetrahedron base and move in an orbit called the reference orbit (**Figure 1**). The mirror satellites form a regular triangle, and the height of the tetrahedron passes through the center of this triangle [10].

For monitoring the fire area in this manuscript, it was assumed that the ground sample distance (GSD) should be equal to 30 m. To achieve this value of GSD, the distance between the plane with the mirror satellites and the imaging satellite, which corresponds

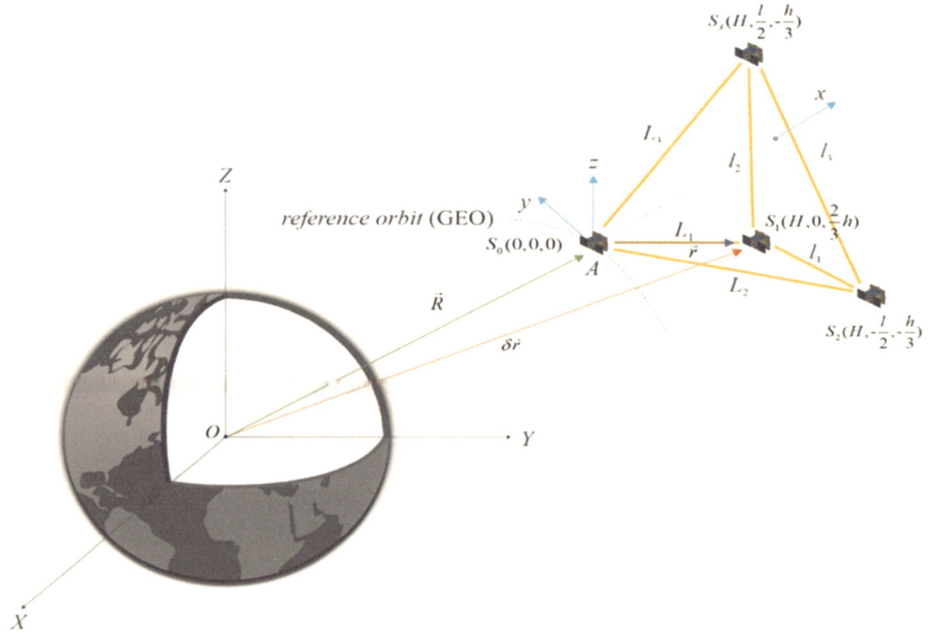

Figure 1.
Tetrahedral satellite formation flying.

to the focal length of the telescope, is assumed to be 21.6 m, and the distance between the mirror satellites, affecting the telescope's aperture, is assumed to be 26.5 m.

The use of distributed satellite systems for aperture synthesis opens up new possibilities for space-based Earth observation. However, this method also brings unique challenges in preserving the geometry of the satellite formation amidst external disturbances. In this paper, gravitational perturbations caused by the Earth's non-sphericity and inhomogeneous mass distribution, represented by the J_2 zonal harmonic, are analyzed in the development of a relative dynamic model for the satellite formation.

In the **Figure 1** are shown following coordinate systems for describing the motion of satellite formation:

OXYZ – inertial coordinate system (ICS), the center of which is located at the Earth's center of mass, the *OZ* axis is directed along the Earth's rotation axis, the *OX* is directed to the point of the Spring equinox of the J2000 epoch.

Axyz – local vertical local horizontal coordinate system (LVLH), the center of which is in the reference satellite represented by imaging satellite, the *Ax* axis is directed along the radius vector of the reference satellite from the center of the Earth, the *Az* axis is normal to the orbit plane in the direction of the orbital momentum, the *Ay* axis completes the system to the right-handed system.

According to [11] considering the inhomogeneity of the Earth's gravitational field represented by J_2 zonal harmonic requires expressing the gravitational potential in the following form:

$$U_2 = -\frac{1}{2}\mu J_2 \frac{r_3^2}{R^3}\left(3\sin^2\varphi - 1\right), \tag{1}$$

where r_3 is the radius of the Earth, R is the radius vector of the reference satellite in the ICS, φ is the latitude of the reference satellite in the ICS.

Further using (1), the equations of satellite motion in formation relative to the reference satellite in LVLH frame were derived in the form [11]:

$$\begin{gathered} \ddot{x}_i - 2nc\dot{y}_i - \left(5c^2 - 2\right)n^2 x_i = u_x, \\ \ddot{y}_i + 2nc\dot{x}_i = u_y, \\ \ddot{z}_i + q_i^2 z_i = 2f_i q_i \cos\left(q_i t + \varphi\right) + u_z, \end{gathered} \tag{2}$$

$$\begin{gathered} n = \sqrt{\frac{\mu}{r_E^3}}, c = \sqrt{1+s}, s = \frac{3J_2 r_E^2}{8R^2}(1 + 3\cos 2I), \\ q_i = nc - \left(\cos\gamma_i \sin\gamma_i \cot\Omega_i - \sin^2\gamma_i \cos I_{1i}\right)\left(\dot{\Omega}_{1i} - \dot{\Omega}_{2i}\right) \\ -\dot{\Omega}_{1i}\cos I_{1i}, \end{gathered} \tag{3}$$

$$\gamma_i = \cot^{-1}\left(\frac{\cot I_{2i}\sin I_{1i} - \cos I_{1i}\cos\Omega_i}{\sin\Omega_i}\right), \Omega_i = \frac{z_i(t_0)}{R\sin I} \tag{4}$$

$$\begin{gathered} I_{1i} = \frac{v_{iz}(t_0)}{kR} + I_{2i}, I_{2i} \approx I, k = nc + \frac{3nJ_2 r_E^2}{2R^2}\cos^2 I, \\ \dot{\Omega}_{1i} = -\frac{3nJ_2 r_E^2}{2R^2}\cos I_{1i}, \dot{\Omega}_{2i} = -\frac{3nJ_2 r_E^2}{2R^2}\cos I_{2i}, \end{gathered} \tag{5}$$

$$\begin{gathered} f_i = -\frac{\sin I_{1i}\sin I_{2i}\sin\Omega_i}{\sin\Phi_i}\left(\dot{\Omega}_{1i} - \dot{\Omega}_{2i}\right)R, \\ \Phi_i = \cos^{-1}\left(\cos I_{1i}\cos I_{2i} + \sin I_{1i}\sin I_{2i}\cos\Omega_i\right), i = 1\ldots3. \end{gathered} \tag{6}$$

where x, y, z is the position of satellite in LVLH (position of imaging satellite), I is the inclination of orbit of reference satellite, I_2 is the inclination of the deputy satellites in formation represented by mirror satellites, $z_i(t_0)$ is the initial value for coordinate z in relative position of satellites in formation, $v_{iz}(t_0)$ the initial value of the velocity on the z axis, R is the radius-vector of reference satellite in ICS, i is the deputy satellite number, u_x, u_y, u_z are the control forces.

To determine the deviation of the current geometrical configuration from the tetrahedral one, we introduced the volume of the tetrahedron [12]:

$$V_T = \frac{1}{6} det\left| \overrightarrow{L_1}, \overrightarrow{L_2}, \overrightarrow{L_3} \right|, \tag{7}$$

where $\overrightarrow{L_1} = [x_1, y_1, z_1]$, $\overrightarrow{L_2} = [x_2, y_2, z_2]$, $\overrightarrow{L_3} = [x_3, y_3, z_3]$ are relative distances between the imaging satellite and mirror satellites in LVLH.

We are focused on developing a control system for a tetrahedral satellite formation in geostationary orbit, designed to achieve high temporal and spatial resolution for monitoring forest fires. This formation acts as a synthetic aperture telescope working in the infrared spectrum. The main goal of the control system is to ensure that the formation maintains its precise geometric configuration with micrometer accuracy. Specifically, the system must keep the relative positioning error of the mirror satellites in the tetrahedral structure within the range of 0.1 to 1 μm.

The equations describing the controlled motion of a satellite within the formation are given by (2), where the control forces are assumed to be linear functions:

$$\begin{aligned} u_x &= -k_x(x_T - x) - k_{Vx}v_x, \\ u_y &= -k_y(y_T - y) - k_{Vy}v_y, \\ u_z &= -k_z(z_T - z) - k_{Vz}v_z, \end{aligned} \tag{8}$$

where x_T, y_T, z_T are required position of satellite in formation relative the reference coordinate system, v_x, v_y, v_z are relative velocity of satellites in formation, $k_x, k_{Vx}, k_y, k_{Vy}, k_z, k_{Vz}$ are constant coefficients.

To determine the unknown feedback coefficients $k_x, k_{Vx}, k_y, k_{Vy}, k_z, k_{Vz}$ in the control acceleration expressions (10), we applied several approaches from linear control theory:

1. the root locus method (RLM), which provides the desired response speed, damping, and system stability;

2. the linear quadratic regulator (LQR), which minimizes state error and control effort;

3. H_2 optimal control, which reduces the system's sensitivity to disturbances;

4. H_∞ optimal control, which ensures the robustness of the system to disturbances;

5. mixed H_∞ control offering a flexible balance between performance and robustness.

Let us consider each approach in more detail.

2.1 Root method (RLM)

At the first stage we considered the classical approach for the root method providing the special location of the closed-loop poles of the system in the complex plain.

As can be seen from (2) the second equation of the system does not contain the variables included in the first and second equations, then it can be considered individually. We will look for a solution to the equation in the form:

$$y = Be^{\lambda t}. \tag{9}$$

From here, we obtain an algebraic second-order characteristic equation:

$$\lambda^2 + k_{Vy}\lambda + q^2 + k_y = 0. \tag{10}$$

Consider the first and third equations of system (2). We will look for their solutions in the form:

$$x = Ae^{\lambda t}, z = Ce^{\lambda t}. \tag{11}$$

From here, we obtain the corresponding system of algebraic equations:

$$\begin{gathered} A\left(\lambda^2 + k_{Vx}\lambda + k_x - \left(5n^2c^2 - 2n^2\right)\right) - 2nc\lambda C = 0, \\ C\left(\lambda^2 + k_{Vy}\lambda + k_y\right) + 2nc\lambda A = 0. \end{gathered} \tag{12}$$

Since the system of Eq. (12) must have a non-zero solution for A and C, the determinant of this system is equal to zero:

$$\begin{vmatrix} \lambda^2 + k_{Vx}\lambda + k_x - (5n^2c^2 - 2n^2) & -2nc\lambda \\ 2nc\lambda & \lambda^2 + k_{Vy}\lambda + k_y \end{vmatrix} = 0 \tag{13}$$

Computing the determinant in (13), further, collecting the terms in the obtained expression by $\lambda^4, \lambda^3, \lambda^2, \lambda,$ and equating this expression to zero, we obtain the algebraic fourth-order characteristic equation in the form:

$$\begin{aligned} \lambda^4 + \lambda^3\left(k_{Vx} + k_{Vy}\right) + \lambda^2\left(k_x + k_y + k_{Vx}k_{Vy} - \left(5n^2c^2 - 2n^2\right) - 4n^2c^2\right) \\ +\lambda\left(k_x k_{Vy} + k_{Vx}k_y - k_{Vy}\left(5n^2c^2 - 2n^2\right)\right) + k_x k_y \\ -\left(5n^2c^2 - 2n^2\right)k_y = 0 \end{aligned} \tag{14}$$

We define the required roots of the second-order characteristic equation as the roots of the Butterworth polynomial since it provides maximally flat and monotonic transition process [13, 14] in the form:

$$\lambda^2 + 1.414\Omega_r\lambda + {\Omega_r}^2 = 0, \tag{15}$$

where $\Omega_r = \frac{t_H}{t_p}$, t_H is the normalized time of the transient process, t_p is the real time of the transient process.

And the roots of the characteristic equation of the fourth order as the roots of the Butterworth polynomial in the form:

$$\lambda^4 + 2.613\Omega_r\lambda^3 + 3.4141\Omega_r{}^2\lambda^2 + 2.613\Omega_r{}^3\lambda + \Omega_r{}^4 = 0 \tag{16}$$

The roots of Butterworth polynomial locate on the circle of radius Ω_r in the left half of complex plain.

Further equating in formulas (20) and (15) terms at the similar power of λ we obtain expressions for determining the unknown feedback coefficients:

$$k_{Vy} = 1.414\Omega_r, k_y = \Omega_r{}^2 - q^2. \tag{17}$$

Equating in formulas (14) and (16) terms at the similar power of λ we obtain a system of algebraic equations for determining the unknown feedback coefficients:

$$\begin{gathered} k_{Vx} + k_{Vy} = 2.613\Omega_r, \\ k_x + k_y + k_{Vx}k_{Vy} - \left(5n^2c^2 - 2n^2\right) - 4n^2c^2 = 3.4141\Omega_r{}^2 \\ k_xk_{Vy} + k_{Vx}k_y - k_{Vy}\left(5n^2c^2 - 2n^2\right) = 2.613\Omega_r{}^3, \\ k_xk_y - \left(5n^2c^2 - 2n^2\right)k_y = \Omega_r{}^4. \end{gathered} \tag{18}$$

At the following stage we considered another approach for the root method proving the location of the closed-loop poles of the system in LMI regions.

LMI regions are convex subsets D of the complex plane characterized by [14]:

$$D = \left\{z \in C : L + Mz + M^T\bar{z} < 0\right\}, \tag{19}$$

where $M = \left\{\mu_{ij}\right\}_{1\le i,j\le m}$ and $L = L^T = \left\{\lambda_{ij}\right\}_{1\le i,j\le m}$ are fixed real matrices. The matrix-valued function

$$f_D(z) := L + Mz + M^T\bar{z} \tag{20}$$

is called characteristic function of the region D.

The approach of control synthesis as the function $\vec{u} = K\vec{X}$ that provides the robust pole placement in LMI regions usually is considered for the linear systems of the type:

$$\begin{aligned} \dot{\vec{X}} &= A\vec{X} + B_1\vec{w} + B_2\vec{u}, \\ \vec{Z} &= C_1\vec{X} + D_{11}\vec{w} + D_{12}\vec{u}, \\ \vec{Y} &= C_2\vec{X} + D_{12}\vec{w} + D_{22}\vec{u}, \end{aligned} \tag{21}$$

where $\vec{X}$ is the state vector of the system, $\vec{w}$ is the exogenous input, $\vec{u}$ is control, $\vec{Z}$ is the regulated output, $\vec{Y}$ is the measured output.

The closed-loop poles of (21) lie in the LMI region (19) if and only if there exists a symmetric positive definite matrix P satisfying [14]:

$$\begin{gathered} \left[\lambda_{ij}P + \mu_{ij}(A + B_2K)P + \mu_{ij}P + \mu_{ji}P(A + B_2K)^T\right]_{1\le i,j\le m} < 0 \\ P > 0 \end{gathered} \tag{22}$$

with the notation $[S_{ij}]_{1\le i,j\le m} = \begin{pmatrix} S_{11} & \cdots & S_{i1} \\ \vdots & \ddots & \vdots \\ S_{m1} & \cdots & S_{mm} \end{pmatrix}$..

LMI region can be represented individually or as intersection of several LMI regions. In this chapter, this region is given as the intersection of a disk of radius r with origin at the point $(-h, 0)$ and a conical sector with inner angle θ:

$$f_D(z) := \begin{pmatrix} -r & \bar{z}+h \\ z+h & -r \end{pmatrix}, \tag{23}$$

$$f_D(z) := \begin{pmatrix} sin\,\frac{\theta}{2}(z+\bar{z}) & -cos\,\frac{\theta}{2}(z-\bar{z}) \\ cos\,\frac{\theta}{2}(z-\bar{z}) & sin\,\frac{\theta}{2}(z-\bar{z}) \end{pmatrix}. \tag{24}$$

The matrices of Eq. (21) used for synthesis of controller has the following form:

$$A = \begin{bmatrix} 0 & 0 & 0 & 1 & 0 & 0 \\ 0 & 0 & 0 & 0 & 1 & 0 \\ 0 & 0 & 0 & 0 & 0 & 1 \\ (5c^2-2)n^2 & 0 & 0 & 0 & 2nc & 0 \\ 0 & 0 & 0 & -2nc & 0 & 0 \\ 0 & 0 & -q^2 & 0 & 0 & 0 \end{bmatrix}, B_1 = \begin{bmatrix} 0 & 0 & 0 \\ 0 & 0 & 0 \\ 0 & 0 & 0 \\ 0 & 0 & 0 \\ 0 & 0 & 0 \\ 0 & 0 & 0 \end{bmatrix}, B_2 = \begin{bmatrix} 0 & 0 & 0 \\ 0 & 0 & 0 \\ 0 & 0 & 0 \\ 1 & 0 & 0 \\ 0 & 1 & 0 \\ 0 & 0 & 1 \end{bmatrix} \tag{25}$$

$$C_1 = diag[1,1,1,1,1,1], C_2 = diag[1,1,1,1,1,1], \tag{26}$$

$$D_{11} = 0_{6\times3}, D_{12} = [0_{3\times3}; diag[1,1,1]], D_{21} = 0_{6\times3}, D_{22} = 0_{6\times3} \tag{27}$$

2.2 Linear quadratic regulator (LQR)

The second method to synthesize the controller for keeping the satellite formation with high accuracy is a linear quadratic regulator (LQR), which is obtained by minimizing the quality criterion of the form [15]:

$$J = \frac{1}{2}\int\left(\overrightarrow{\Delta r}^T Q \overrightarrow{\Delta r} + \vec{u}^T R \vec{u}\right) dt \tag{28}$$

where $\overrightarrow{\Delta r} = [x_T - x, y_T - y, z_T - z, v_x, v_y, v_z]$, Q, R are positive matrices with constant components, x_T, y_T, z_T - is the required position of the satellite in a formation.

The control acceleration $\vec{u} = [u_x, u_y, u_z]$ for the linear system in the form (23) obtained as result of minimizing the performance criterion (28) has the form [15]:

$$\vec{u} = -R^{-1}B^T P \overrightarrow{\Delta r} = K\overrightarrow{\Delta r}, \tag{29}$$

where $B = B_2$ introduced in (27), the matrix R has the form:

$$R = 0.0001 \cdot diag[1,1,1]. \tag{30}$$

And the matrix P can be determined from the equation:

$$A^T P + PA - PBR^{-1}B^T P + Q = 0, \tag{31}$$

where $A, B = B_2$ are the matrices introduced in (25), the matrix Q has the form:

$$Q = 0.0001 \cdot diag[1, 1, 1, 1, 1, 1]. \tag{32}$$

2.3 H_2 optimal control

H_2 optimal controller is considered for the linear systems with a state space representation similar to (23):

$$\begin{aligned} \dot{\vec{X}} &= A\vec{X} + B_1\vec{w} + B_2\vec{u}, \\ \vec{Z} &= C_1\vec{X} + D_{12}\vec{u}, \\ \vec{Y} &= C_2\vec{X} + D_{12}\vec{w} + D_{22}\vec{u}, \end{aligned} \tag{33}$$

where $D_{12}^T D_{12} > 0$.

The state feedback synthesis control problem reduces to developing a controller $\vec{u} = K\vec{X}$ that minimizes the H_2 norm of the transfer function $H_{zw}(s)$ [16]:

$$\|H_{zw}(s)\|_2 = \sqrt{\frac{1}{2\pi}\int_{-\infty}^{\infty} trace\left(H_{zw}(j\omega)^T H_{zw}(j\omega)\right)d\omega} \tag{34}$$

To calculate this norm we need to solve the following problem [16]:

$$\begin{aligned} &\min trace\left[(C_1 + D_{12}K)P(C_1 + D_{12}K)^T\right] \\ &s.t.(A + B_2K)P + P(A + B_2K)^T + B_1B_1^T \le 0, \\ &P = P^T > 0 \end{aligned} \tag{35}$$

Introducing $X_2 = X_2^T = P$, $L = KP$, $Q = Q^T$ and using appropriate transformations (35) is equivalent to LMI problem:

$$\begin{aligned} &\min{}_{Q,X_2,L} trace(Q) \\ &s.t. AX_2 + X_2A^T + B_2L + L^TB_2^T + B_1B_1^T \le 0, \\ &\begin{bmatrix} X_2 & X_2C_1^T + L^TD_{12}^T \\ C_1X_2 + D_{12}L & Q \end{bmatrix} \ge 0. \end{aligned} \tag{36}$$

Using the matrices A, B_1, B_2, C_1, D_{12} introduced in (25), (26), (28) and finding (X_2, L) satisfying (36) the state feedback gain matrix is calculated as $K = LX_2^{-1}$..

2.4 H_∞ optimal control

H_∞ optimal controller is considered for the linear systems with a state space representation similar to (21). The state feedback synthesis control problem reduces to developing a controller $\vec{u} = K\vec{X}$ that minimizes the H_∞ norm of the transfer function $H_{zw}(s)$ [17]:

$$\|H_{zw}(s)\|_\infty = \sup_{\omega \in R} \sigma_{max}(F(j\omega)) \tag{37}$$

To calculate this norm, we need to solve the following problem [17]:

$$\begin{gathered}\min_{P,K,\gamma} \gamma \\ s.t. \begin{bmatrix} (A + B_2K)^T P + P(A + B_2K) & PB_1 & (C_1 + D_{12}K)^T \\ B_1^T P & -\gamma I & D_{11}^T \\ C_1 + D_{12}K & D_{11} & -\gamma I \end{bmatrix} < 0, \\ P > 0.\end{gathered} \tag{38}$$

Introducing $X_\infty = X_\infty^T = P, L = KP$ and using appropriate transformations (38) is equivalent to:

$$\begin{gathered}\min_{X_\infty,L,\gamma} \gamma \\ s.t. \begin{bmatrix} (AX_\infty + B_2L)^T + (AX_\infty + B_2L) & B_1 & (C_1 + D_{12}L)^T \\ B_1^T & -\gamma I & D_{11}^T \\ C_1 + D_{12}L & D_{11} & -\gamma I \end{bmatrix} < 0, \\ X_\infty > 0.\end{gathered} \tag{39}$$

Using the matrices A, B_1, B_2, C_1, D_{12} introduced in (25)–(27), and finding (X_∞, L) satisfying (39) the state feedback gain matrix is calculated as $K = LX_\infty^{-1}$..

2.5 Mixed H_2/H_∞ control

Mixed H_2/H_∞ control is considered for the linear systems with a state space representation:

$$\begin{aligned} \dot{\vec{X}} &= A\vec{X} + B_1\vec{w} + B_2\vec{u}, \\ \overrightarrow{Z_\infty} &= C_\infty\vec{X} + D_{\infty 1}\vec{w} + D_{\infty 2}\vec{u}, \\ \overrightarrow{Z_2} &= C_2\vec{X} + D_{21}\vec{w} + D_{22}\vec{u}, \\ \vec{Y} &= C_y\vec{X} + D_{y1}\vec{w}, \end{aligned} \tag{40}$$

where $\vec{X}$ is the state, $\vec{u}$ is the control, $\vec{Y}$ is the output, where $\vec{w} \to \overrightarrow{Z_\infty}$ is the H_∞ performance channel, $\vec{w} \to \overrightarrow{Z_2}$ is the H_2 performance channel.

Mixed H_2/H_∞ output feedback controller for the plant (40) can be obtained in the form [18]:

$$\begin{aligned} \dot{\overrightarrow{x_k}} &= A_k\overrightarrow{x_k} + B_k\vec{Y}, \\ \vec{u} &= C_k\overrightarrow{x_k} + D_k\vec{Y}. \end{aligned} \tag{41}$$

This controller stabilizes the system exponentially, the H_∞ performance channel has a performance level $\left\|H_{\vec{w}\to\overrightarrow{z_\infty}}(K)\right\|_\infty \leq \gamma$, the H_2 performance $\left\|H_{\vec{w}\to\overrightarrow{z_2}}(K)\right\|_2$ is minimized among all K stabilizing the system exponentially and having prescribed H_∞ performance level γ.

Thus, the mixed H_2/H_∞-synthesis problem is the optimization program [18]:

$$\begin{gathered} \min \left\|H_{\vec{w}\to\overrightarrow{z_2}}(K)\right\|_2 \\ s.t. \left\|H_{\vec{w}\to\overrightarrow{z_\infty}}(K)\right\|_\infty \leq \gamma, \\ K \textit{ stabilizes system internally}, \end{gathered} \tag{42}$$

where $H_{\vec{w}\to\overrightarrow{z_2}}(K,s)$ denotes the transfer function of the H_2 closed-loop performance channel, $H_{\vec{w}\to\overrightarrow{z_\infty}}(K)$ denotes the transfer function of the H_∞ closed-loop performance channel, γ is a threshold.

The solution of presented problem is presented in [18].

The matrices $A, B_1, B_2, C_\infty, D_{\infty 1}, D_{\infty 2}, C_2, D_{21}, D_{22}, C_y, D_{y1}$ used for synthesis of H_2/H_∞ controller has the following form correspondingly:

$$A = \begin{bmatrix} 0 & 0 & 0 & 1 & 0 & 0 \\ 0 & 0 & 0 & 0 & 1 & 0 \\ 0 & 0 & 0 & 0 & 0 & 1 \\ (5c^2-2)n^2 & 0 & 0 & 0 & 2nc & 0 \\ 0 & 0 & 0 & -2nc & 0 & 0 \\ 0 & 0 & -q^2 & 0 & 0 & 0 \end{bmatrix}, B_1 = \begin{bmatrix} 0 & 0 & 0 \\ 0 & 0 & 0 \\ 0 & 0 & 0 \\ 0 & 0 & 0 \\ 0 & 0 & 0 \\ 0 & 0 & 0 \end{bmatrix}, B_2 = \begin{bmatrix} 0 & 0 & 0 \\ 0 & 0 & 0 \\ 0 & 0 & 0 \\ 1 & 0 & 0 \\ 0 & 1 & 0 \\ 0 & 0 & 1 \end{bmatrix} \tag{43}$$

$$C_\infty = diag[1,1,1,1,1,1], C_2 = diag[1,1,1,1,1,1], C_y = diag[1,1,1,1,1,1], \tag{44}$$

$$D_{\infty 1} = 0_{6\times 3}, \quad D_{\infty 2} = [0_{3\times 3}; diag[1,1,1]], \quad D_{21} = 0_{6\times 3}, \quad D_{22} = 0_{6\times 3}, \quad D_{y1} = 0_{6\times 3} \tag{45}$$

Propulsion systems in satellite formation control are vital for executing accurate maneuvers and sustaining the intended configuration of the formation. These systems, which include both chemical and electric propulsion, allow the satellites to perform precise adjustments to maintain the required formation structure.

3. Propulsion systems for satellite formation control

The propulsion systems used in satellite formation control play a crucial role in ensuring precise maneuvering, maintaining, and preserving the designated configuration of the formation. Propulsion systems are divided into chemical and electric types, each with its own advantages and disadvantages.

Chemical engines use the reaction between an oxidizer and fuel to generate thrust. They provide high instantaneous thrust, making them useful for quick adjustments to the position of each satellite in the constellation. However, their fuel resources are limited, which reduces the lifespan of not only the propulsion system but also the entire satellite constellation. Hydrazine engines are simple in design and operate on a single fuel. The advantages of these engines include reliability and ease of control. Bipropellant engines use two components: fuel and an oxidizer. These engines have a higher specific impulse but require more complex fuel delivery systems.

Electric propulsion systems are becoming increasingly popular for small spacecraft due to their high efficiency and economical fuel usage. They generate thrust by using electrical energy to accelerate charged particles (ions), significantly increasing the specific impulse compared to chemical engines. There are several types of electric propulsion systems commonly used for small spacecraft:

Ion engines use an electric field to accelerate ions, allowing them to generate small but constant thrust over extended periods. These engines are highly efficient, especially for long-duration missions. They require minimal fuel but significant power for electricity generation.

Hall-effect thrusters are among the most widely used types of electric propulsion systems. They also use xenon to create thrust with the help of a magnetic field that contains and accelerates ions. This is a simpler and more reliable option compared to ion engines, making it suitable for maneuvering and maintaining the configuration of satellite formation.

Electrothermal engines, such as arcjet or resistojet thrusters, heat the propellant using electrical energy and expel it through a nozzle to generate thrust. These engines provide higher thrust than ion engines but are less efficient in terms of specific impulse.

Since small spacecraft are limited in mass and volume, electric propulsion systems are most commonly used for orbit correction, attitude control, and maintaining the required configuration of satellite constellations. They can operate for extended periods with minimal fuel consumption and high accuracy.

For example, the authors of [19] explored the use of a cold gas propulsion system for precise and rapid control of spacecraft motion. The article describes various control modes required to maintain the constellation's configuration. The small spacecraft UoSAT-12 is equipped with a single N_2O resistojet engine for orbit maintenance and 10 nitrogen cold gas thrusters for attitude control.

Danil Ivanov and his colleagues [20] highlight the widespread adoption of electric propulsion (EP) systems for satellite formation control. Electric propulsion systems, such as ion thrusters and Hall-effect thrusters, are noted for their high efficiency, low fuel consumption, and ability to provide precise thrust adjustments. These systems are ideal for long-duration missions that require gradual changes in satellite positions. Due to their low thrust output, EP systems are typically used for fine-tuning or station-keeping in satellite formations.

Delfi-n3Xt [21], launched in 2013, is the second Delfi 3 U CubeSat developed by Delft University of Technology for educational purposes and technology demonstration. Delfi-n3Xt successfully demonstrated the T3-μPS cold gas micropropulsion technology, developed by TU Delft, TNO, the University of Twente, and SystematIC Design BV. The micropropulsion system uses cold gas generators (CGGs) to ignite a solid fuel pellet and release nitrogen gas, which is then compressed and stored in a tank for thrust generation. This indicates that the system was only pressurized in space prior to thrust generation. The micropropulsion system was capable of delivering 6 mN of thrust while consuming 10 W of power to ignite the CGG for up to 10 seconds. Two ignitions were successfully performed in orbit; however, a malfunction occurred in the ignition system. The technology demonstration mission concluded that it is recommended to incorporate a parallel design for the ignition system to avoid single points of failure and reduce the risk of ignition system malfunction.

In Ref. [22], a new hybrid propulsion system combining Coulomb forces with electric/ion propulsion has been developed for spacecraft formation flying in geostationary (GEO) and other high Earth orbits. The effectiveness of this hybrid propulsion approach is demonstrated through a tetrahedral formation scenario. The APF method ensures safe, collision-free aggregation, while Sliding Mode Control provides robust performance, minimizing the effects of external disturbances like differential solar radiation and internal factors such as fuel consumption-induced mass changes. For the tetrahedral formation in this study, the control effort is reduced thanks to the controller's inherent robustness, resulting in a fuel-efficient system for close proximity formation flying missions.

The authors [23] have developed miniature ion thrusters designed for small spacecraft in the 50 kg class, incorporating a miniature neutralization system. Two types of ion thruster systems were proposed based on their neutralization methods: unipolar and bipolar ion thrusters. Typically, the unipolar thruster offers higher performance, while the bipolar thruster is simpler and lighter. In this study, the performance of these miniature ion thrusters was evaluated in terms of thrust, power consumption, and specific impulse for both unipolar and bipolar modes. Notably, the power and propellant consumption of the neutralizer were included in the performance analysis.

The unipolar thruster generated 379 μN of thrust with a specific impulse of 1410 seconds using 19.9 W of total power, or 297 μN of thrust with a specific impulse of 1100 seconds using 15.1 W of power. In bipolar mode, the thrust and specific impulse were 10–20% lower than in unipolar mode at the same power levels. Despite this, both types of miniature ion thrusters demonstrated performance suitable for 50 kg spacecraft.

The literature review presented above confirms that existing satellite propulsion systems in practice are capable of providing the necessary force to maintain a tetrahedral configuration of satellite formation. As a result, the positive outcomes of the H∞ and H2/H∞ regulators used in this study are evident. Thus, it can be concluded that there are propulsion systems that can provide the required thrust force of approximately 0.19 N for satellites S1, S2, and S3 to maintain the configuration.

Based on the obtained results, it can be concluded that for the successful completion of the mission involving the formation of satellites designed to monitor the Earth's surface, the most suitable control methods are those using H∞ and H2/H∞ regulators. These methods provide the necessary accuracy for mutual positioning of the satellites, ensure high-quality transient processes, and maintain the configuration with high precision. According to the results of modeling and thrust calculations (**Figures 2–4**) using engines with compressed gas and a specific impulse of

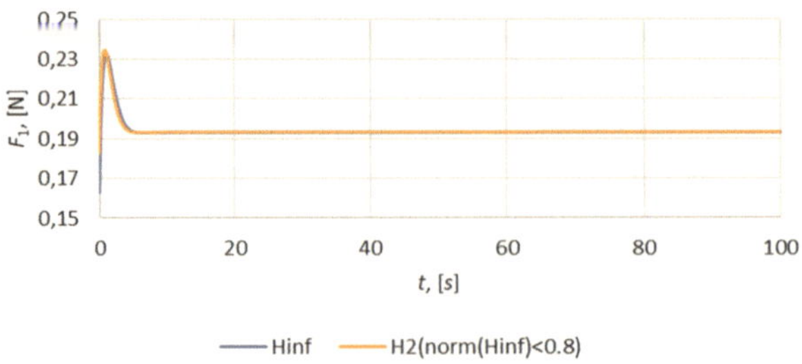

Figure 2.
Thrust force for satellite S1.

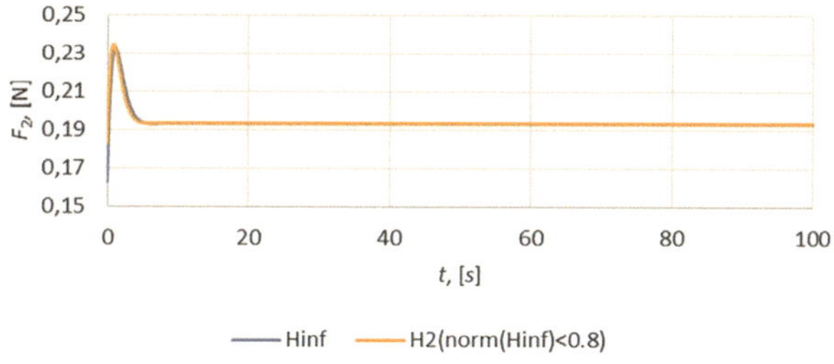

Figure 3.
Thrust force for satellite S2.

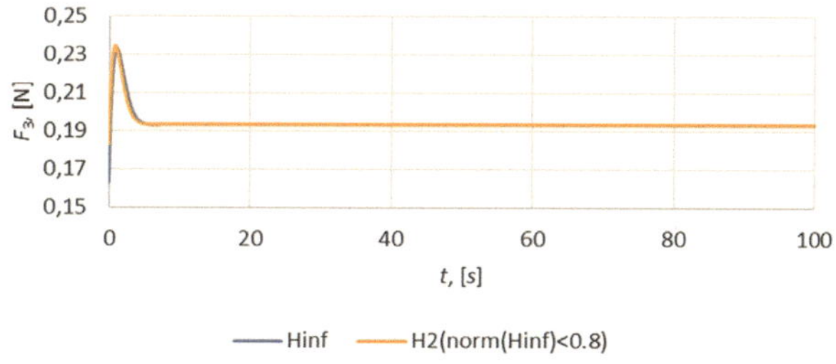

Figure 4.
Thrust force for satellite S3.

70 seconds, the required fuel consumption for the satellite was 9.86 kg per year. This emphasizes the need to improve the controller in order to reduce fuel consumption for satellites in the formation.

4. Conclusion

This study focused on the problem of controlling satellite formations designed for real-time Earth surface monitoring. The proposed control system ensures high precision in maintaining the geometric configuration of the satellites, which is critical for high-resolution tasks such as disaster monitoring. The analysis demonstrated that H∞ and H2/H∞ regulators are the most suitable control methods for achieving accurate satellite positioning and ensuring system stability.

Additionally, various propulsion systems, including both chemical and electric types, were analyzed for maintaining satellite formations. Electric propulsion systems, such as ion and Hall-effect thrusters, proved to be highly efficient, offering precise orbital adjustments while minimizing fuel consumption. The paper also explored the potential of hybrid propulsion systems, combining electric and ion thrusters, as well as the development of miniature ion thrusters for small satellites. These systems are ideal for maintaining satellite configurations with minimal fuel use.

The results indicate that current propulsion technologies are capable of providing the necessary thrust to maintain complex formations, such as tetrahedral structures. Specifically, the control and propulsion systems used in this study can generate approximately 0.19 N of thrust for the satellites, allowing them to sustain their configuration during long-term missions with minimal fuel expenditure. Therefore, the proposed control methods and propulsion systems can be effectively applied to satellite missions that require high resolution and extended operational timeframes.

Author details

Zaure Rakisheva[1*], Nazgul Kaliyeva[1], Nursultan Doszhan[1], Anna Sukhenko[2] and Gulama Garip Alisher Ibrayev[1]

1 Al-Farabi Kazakh National University, Almaty, Kazakhstan

2 Kazakh-British Technical University, Almaty, Kazakhstan

*Address all correspondence to: zaure.rakisheva@kaznu.kz

References

[1] Suzumoto R, Ikari S, Miyamura N, Nakasuka S. Experimental study for μm-class control of relative position and attitude for synthetic aperture telescope using formation flying micro-satellites. Proceedings of the IFAC-PapersOnLine. 2020;**53**:5701-5708. DOI: 10.1016/j.ifacol.2020.12.1597

[2] Henault F. Imaging and nulling properties of sparse-aperture Fizeau interferometers, SPIE astronomical telescopes + instrumentation. In: Proceedings SPIE 9146 Optical and Infrared Interferometry IV, Montréal, Quebec, Canada. Vol. 9146. 2014. pp. 366-379. DOI: 10.1117/12.2055069

[3] Rousset G, Mugnier LM, Cassaing F, Sorrente B. Imaging with multi-aperture optical telescopes and an application. Comptes Rendus De L'Académie Des Sciences - Series IV - Physics. 2001;**2**: 17-25. DOI: 10.1016/s1296-2147(01)01158-1

[4] Mugnier L, Cassaing F, Rousset G, Baron F, Michau V, Mocœur I, et al. Continuous high-resolution earth observation with multiple aperture optical telescopes. In: Proceedings of the OPTRO 2005 International Symposium. Chatillon cedex, France; 2005

[5] Mesrine M, Thomas E, Garin S, Blanc P, Alis C, Cassaing F, et al. High resolution earth observation from geostationary orbit by optical aperture synthesys. In: Proceedings SPIE 10567, International Conference on Space Optics — ICSO 2006. Vol. 10567. Noordwijk, Netherlands: ESA and CNES; 2017. pp. 71-77. DOI: 10.1117/12.2308095

[6] Dolkens D, Kuiper JM. A deployable telescope for submeter resolutions from microsatellite platforms. In: Proc SPIE 10563, International Conference on Space Optics — ICSO 2014. Vol. 10563. Canary Islands, Spain: Tenerife; 2017. pp. 992-1000. DOI: 10.1117/12.2304245

[7] Cui S, Xu B, Luo S, Xu H, Cai Z, Luo Z, et al. Determining topological charge based on an improved Fizeau interferometer, Optics Express. Vol. 27. USA: Optica Publishing Group; 2019. p. 12774. DOI: 10.1364/oe.27.012774

[8] Chenghao Z, Zhile W, Shuqing Z. Numerical simulation of optical synthetic aperture imaging system. Space Optics and Earth Imaging and Space Navigation. Beijing, China: SPIE. digital library; 2017, p.104630V. DOI: 10.1117/12.2282811

[9] Leitner J. Formation flying-the future of remote sensing from space. In: Proceedings of the 18th International Symposium on Space Flight Dynamics (ESA SP-548). USA: SAO/NASA Astrophysics Data System; 2004. p. 621

[10] Kaliyeva N, Rakisheva Z, Doszhan N, Sukhenko A, Ibrayev G-GA, Nakasuka S, et al. Relative dynamics and control of satellite formation flying representing the synthetic aperture telescope on geostationary orbit. Engineered Science. 2024;**29**:1131. DOI: 10.30919/es1131

[11] Xu C, Tsoi R, Sneeuw N. Analysis of J2-perturbed relative orbits for satellite formation flying. International Association of Geodesy Symposia. 2005;**129**:30-35. DOI: 10.1007/3-540-26932-0_6

[12] Liu A, Joe B. On the shape of tetrahedra from bisection. Mathematics of Computation. 1994;**63**:141. DOI: 10.2307/2153566

[13] Ellis G. Control System Design Guide Chapter 9 Filter in Control Systems.

4th ed. Amsterdam, Netherland: Elsevier Inc.; 2012

[14] Chilali M, Gahinet P, Apkarian P. Robust pole placement in LMI regions. IEEE Transactions on Automatic Control. 1999;**44**:2257-2270. DOI: 10.1109/9.811208

[15] Anderson BDO, Moore JB. Optimal Control: Linear Quadratic Methods. Dover Publications; 2007. pp. 130-149

[16] Zolotas AC, Korba P, Chaudhuri B, Jaimoukha IM. $\mathscr{H}2$ LMI-based robust control for damping oscillations in power systems. In: Proceedings of the 2007 IEEE International Conference on System of Systems Engineering. SOSE; 2007. pp. 1-8

[17] Gahinet P, Apkarian P. A linear matrix inequality approach to H∞ control. International Journal of Robust and Nonlinear Control. 1994;**4**:421-448. DOI: 10.1002/rnc.4590040403

[18] Apkarian P, Noll D, Rondepierre A. Mixed H2/H∞ control via nonsmooth optimization. SIAM Journal on Control and Optimization. 2008;**47**:6460-6465. DOI: 10.1137/070685026

[19] Steyn WH, Hashida Y. In-orbit attitude and orbit control commissioning of UoSAT-12. Spacecr. Guid. Navig. Control Syst. 2000;**425**:95

[20] Biktimirov S, Ivanov D, Pritykin D. A satellite formation to display pixel images from the sky: Mission design and control algorithms. Advances in Space Research. 2022;**69**:4026-4044. DOI: 10.1016/j.asr.2022.03.018

[21] Guo J, Bouwmeester J, Gill E. In-orbit results of Delfi-n3Xt: Lessons learned and move forward. Acta Astronautica. 2016;**121**:39-50

[22] Saaj CM, Lappas V, Schaub H, Izzo D. Hybrid propulsion system for formation flying using electrostatic forces. Aerospace Science and Technology. 2010;**14**:348-355

[23] Koizumi H, Kuninaka H. Performance evaluation of a miniature ion thruster μ1 with a unipolar and bipolar operation. In: Proc. 32nd International Electric Propulsion Conference. Wiesbaden, Germany: Electric Rocket Propulsion Society; 2011. pp. 1-10